W9-DIY-889

connecting *the* drops

connecting *the* drops

A CITIZENS' GUIDE TO PROTECTING WATER RESOURCES

Karen Schneller-McDonald

Comstock Publishing Associates
a division of
Cornell University Press
ITHACA AND LONDON

A companion website for this book is available at www.thewetnet.net.

First published 2015 by Cornell University Press
First printing, Cornell Paperbacks, 2015

Printed in the United States of America

Library of Congress Cataloging-in-Publication Data

Schneller-McDonald, Karen, author.
    Connecting the drops : a citizens' guide to protecting water resources / Karen Schneller-McDonald.
        pages cm
    Includes bibliographical references and index.
    ISBN 978-0-8014-5310-6 (cloth : alk. paper) —
    ISBN 978-1-5017-0028-6 (pbk. : alk. paper)
    1. Water—Pollution—Environmental aspects. 2. Water-supply—Environmental aspects. 3. Water quality management. 4. Water conservation. 5. Hydrology. 6. Environmentalism. I. Title.
    TD420.S36 2015
    333.91—dc23   2015006283

Cornell University Press strives to use environmentally responsible suppliers and materials to the fullest extent possible in the publishing of its books. Such materials include vegetable-based, low-VOC inks and acid-free papers that are recycled, totally chlorine-free, or partly composed of nonwood fibers. For further information, visit our website at www.cornellpress.cornell.edu.

Cloth printing   10 9 8 7 6 5 4 3 2 1

Paperback printing   10 9 8 7 6 5 4 3 2 1

For my father,
Alfred Schneller,
who planted the seeds

Water . . . reaches everywhere; it touches the past and prepares the future; it moves under the poles and wanders thinly in the heights of air. It can assume forms of exquisite perfection in a snowflake or strip the living to a single shining bone cast up by the sea.

Loren Eiseley, *The Immense Journey*

contents

acknowledgments

This project was very much a joint effort. Many thanks first to my husband, Mac, for his unflagging support, bringing me coffee to keep me going in the morning and wine to relax with in the evening; providing a ready ear for my ramblings through this material; patience for all the time it took; and constructive chapter feedback. Next, to my son, Matt, the English major and my fellow writer who supplied critiques and edits and provided unfailing encouragement for this project. Special thanks to my editor, Susan Schmidt of the Beaufort Writers Group, for her encouragement and unremitting insistence on better writing. And thanks to Kitty Liu at Cornell University Press for her patience and encouragement during the development of this project.

Thanks also to Frank Margiotta for very helpful reviews of the biology content, Angela Sisson, for her exemplary work on maps and figures, and Stephanie LaRose Lewison for topographic map design and edits.

My circle of thanks grows to include the individuals with whom I've worked over the years who have inspired me by their motivation to protect treasured places and their perseverance and determination to take action and prevail.

And my family, from my mother, who reminds me of my roots, to my grandchildren, who remind me of the wonder of discovery. Last but not least, to the inhabitants of the pond where I lost my boots in the mud so long ago; the frogs, voiceless to defend their habitat but certainly not silent when it comes to celebrating a summer evening. They never fail to inspire me.

Introduction

Eventually all things merge into one, and a river runs through it. The river was cut by the world's great flood and runs over rocks from the basement of time. On some of the rocks are timeless raindrops.

NORMAN MACLEAN, *A River Runs Through It*

We face a water crisis. We don't experience this crisis as one sudden catastrophe that dries up wells or poisons drinking water overnight. Instead, we experience insidious, incremental events—oil spills, floods, polluted runoff, hydrofracking operations—that threaten our water and our communities in one place at a time. Small regional disasters accumulate into an increasingly serious and widespread threat to all our streams, rivers, lakes, and groundwater.

Our headlong rush to economic development without heed to the effects on the environment is causing an array of problems, including flooding, pollution, ecological damage, and dwindling water supply. How does this affect you?

Water means more than a scenic photo. It is essential for our health, our families, and our future. Everything we eat and drink is impacted by water. Without a healthy water cycle, life would not exist. If we care about advancing medical technologies to keep ourselves healthy, it makes no sense to take water—our lifeblood—for granted.

We need to protect our water resources. To be effective, we must understand how they are threatened and identify cause-and-effect connections between land use activities, natural water systems, and the water we use and enjoy. This book is about making such connections. It will help you

- protect and improve your community's water by recognizing the natural systems that supply, cleanse, and store water;
- understand how construction projects, natural gas development, and other land use activities affect water resources;
- uncover the facts in news items or local controversies about development projects, hydrofracking, flooding, and water contamination; and
- become an effective advocate for taking care of your community's water resources now and into the future.

We need both a healthy environment that sustains our personal and community health and vibrant, sustainable economic development that does not destroy the benefits we derive from nature. Our ability to accomplish this depends on how well we can "connect the drops."

Environmental Protection

Earth Day 1970. I was poised to change the world—or to at least clean it up and restore it to its natural state. The mood across the country was optimistic, as a *Washington Post* article described: "A great outpouring of Americans—several million in all likelihood—demonstrated yesterday their practical concern for a livable environment on this earth. . . . So many politicians, in fact, took part in yesterday's Earth Day activities that the United States Congress shut down. Scores of senators and congressmen fanned out across the country to appear at rallies, teach-ins and street demonstrations."[1]

That year, my father gave me a copy of John and Mildred Teal's *Life and Death of the Salt Marsh*, with this inscription: "With the hope that you—and other young minds and muscles—can help to swing the pendulum back." He meant the swing of the pendulum from the death of natural systems, like the salt marsh, back to life. I remember thinking—*Sure, I can do that*. And it did seem entirely possible: Congress adopted the National Environmental Policy Act, followed by the Clean Water Act (1972) and the Endangered Species Act (1973). States signed new environmental departments into law. Earth Day seemed on its way to becoming a national phenomenon.

In Maine, a 1966 report estimated the Kennebec River needed a $45 million pollution cleanup. Citizens who wanted to improve the river's condition presented a strong case to the legislature and were willing to fight for it. State legislators responded to the people and their priorities and appropriated funding.[2] As a result, Maine communities improved water quality in the Kennebec River—from what was once described as an "open sewer" to a system that supported aquatic life—with improved access, recreation opportunities, and economic health. The cleanup of the Kennebec allowed people to realize that, in turn, the river sustained them.[3] As a college freshman, I joined that movement to clean up the Kennebec. The project whetted my enthusiasm for environmental protection and led me to believe that, with enough hard work, anything is possible.

Fast forward to 2013. The books on my shelf—*Nature in Fragments*, edited by Elizabeth Johnson and Michael Klemens, *Last Child in the Woods* by Richard Louv, *An Inconvenient Truth* by Al Gore—remind me that the pendulum isn't swinging back but is instead moving farther from environmental protection. Conservation of the earth's resources remains an elusive target.

The optimism of Earth Day in 1970 has been tough to maintain. Faced with cynicism about whether we can protect our environment at all—and cynicism doesn't make anything better—we need to become water-quality advocates and improve our effectiveness in demanding environmental protection. A few ideas emerge:

- *A strong grassroots approach to environmental-protection issues.* We can no longer depend on the "top-down" protection that our major environmental legislation once seemed to promise. Elected officials need support—and sometimes a great prod—to make decisions that favor protection of natural resources at the local level.
- *A view of natural resources in terms of systems.* In nature, living things are connected with their chemical and physical environment and with each other. They move through cycles. We are connected to these natural systems too; they sustain us. To identify how land-use activities and projects affect us, we need to understand how natural systems work— and in turn how changes to those systems affect us.
- *A detailed correlation between land-use activities and their impacts on specific natural resources.* We can mitigate adverse effects of human activities more effectively if we identify specific activities and their effect on plants, animals and the physical and chemical environment. For example, construction in an undeveloped area often affects water quality. To address this we need more specific information. Which activities are causing water pollution? What is the pollutant (for example, chemical, eroded soil, or other material)? What path does it follow through the environment? How does it affect water chemistry, aquatic plants, insects, or fish? Does it pose a health threat to people?
- *A plan for interpreting this information and communicating its significance.* We need to translate scientific or technical information (about both land-use activities and the workings of nature) into terms that make it easier to understand cause and effect connections between impacts and resources. Then we can use this information as a basis for local environmental protection.

Connecting the Drops explores these ideas and develops them into guidelines for improving environmental protection. The need for action is urgent, as recent threats to the environment challenge protection of our land and water in communities across the country.

Conditions that exacerbate these threats include the following:

- the lack of effective regulatory and legal protection;
- legislative attempts to weaken existing environmental regulations;
- reductions in federal and state environmental-protection funding and staff; and
- promotion of energy development without consideration of the true costs to the environment and human health.

For example, these conditions have culminated in the recent controversy over whether to allow hydrofracking for natural gas in New York State. In 2011, the New York State Department of Environmental Conservation, charged with both protecting the environment and promoting the development of mineral resources (including oil and natural gas), produced the lengthy Draft Supplemental Generic Environmental Impact Statement to evaluate the impacts of allowing hydrofracking to proceed in the state. After four months of public review, including numerous public hearings, the state received more than sixty thousand comments, many of them pointing out deficiencies in the impact analysis. During that same year, natural gas companies spent more than $3.2 million lobbying the New York State government. National energy companies advertised heavily in an effort to convince the public that hydrofracking is safe and economically beneficial.

As the energy companies revved their engines across the border in Pennsylvania, where "fracking" is already in progress, many New York State residents united to oppose this form of energy development in their communities. As of November 2013, 175 municipalities had passed bans or moratoriums; eighty-eight more were considering them.[4] In December 2014, after the release of a lengthy public health review of hydrofracking, New York's governor Andrew Cuomo announced his decision to ban the practice in the state.[5]

Although this decision was the first ban by a state with significant natural gas resources, at least twenty other states host active fracking operations. Even with a ban, New York State is still beset by environmental protection issues associated with natural gas infrastructure, storage, and transportation.

While the news from New York is encouraging, natural gas development remains highly controversial. Oil and gas companies insist that hydrofracking is safe, doesn't harm the environment, and will pump much-needed money into local economies. Some local residents support this view, but many scientists, environmental organizations, and local municipalities, along with other residents, believe that it's not possible to conduct hydrofracking safely, no matter how rigorous the regulations. Many communities remain bitterly divided. The oil and gas industry promises jobs and economic growth and assures us of safety, while folks who live next to existing facilities experience adverse health effects, contaminated water, and reduced property values.

This controversy brings into sharp relief the state of natural resource protection today. Conflicts arise as we weigh the issues: the concern of local residents for health, community, and environmental protection; the need for jobs and energy; the reality of scientific research; and the power of corporate interests. As a result, effective protection of natural resources, given a strong start by landmark national legislation in the 1970s, has been left hanging. The task of shoring up environmental protection has become a local grassroots responsibility.

Why Water?

If our entire environment is at risk, why does this book focus on water?

Water is everything. If you doubt that, try living without it for a few days. Whether you live in a city, a small town, or out in the country, when was the last time you thought about where that water in your glass comes from? Why do places that rarely flooded thirty years ago now flood almost every time it rains? Why can't you eat the fish from that pristine-looking river or lake? What we do on the land affects water. Even air pollutants from land use activities find their way into water.

Natural systems provide the water we use, along with other services that benefit our communities. When human activities and pollution change natural systems, we lose those benefits. Sometimes we pay to replace them; sometimes they can't be replaced.

We are part of a system of interconnected water resources. People love water! Any puddle big enough to splash in attracts little kids like a magnet. We record the sound of waterfalls and streams and bring aquariums and bubbling fountains into our homes and businesses. The calming effect of water splashing on rocks and trickling through pebbles eases the stress of our busy lives. We create water parks in our cities and fern-fringed pools in our gardens. Water concerns everyone, even if we have different ideas about how to protect it. Everyone wants clean, abundant water to drink and neighborhoods free from the ravages of flooding.

But dire headlines point to ongoing problems: "Lake Erie Algae a Threat to Ohio Drinking Water";[6] "Louisiana Agency Sues Dozens of Energy Companies for Damage to Wetlands";[7] "Wells Dry, Fertile Plains Turn to Dust."[8]

Recent research findings are no more encouraging:

- Amphibian populations are declining across the United States. A study from Pennsylvania and Oregon state universities documents the significant loss of amphibians, including common species. Amphibians are valuable indicators of environmental stress, including degraded water quality. Declines are attributed to habitat loss, increased ultraviolet radiation, introduced nonnative species including pathogens, and contaminants.[9]
- Recent studies correlate an increased risk of injury and property damage from floods, droughts, and extreme storms with the mismanagement of land and water resources. The Partners for Resilience program, integrating ecosystem management into disaster aid work, documents that wetland loss and degradation, in particular, make some communities more vulnerable to the effects of extreme weather events.[10] In the United States, more than half our original wetland acreage has been drained or converted to other uses. "The increase in flood damages, drought damages, and the declining bird populations are, in part, the result of wetlands degradation and destruction."[11]
- Today's worst watershed stresses may become the "new normal." A study from the Cooperative Institute for Research in Environmental Sciences provides an overview of contemporary annual water demand and compares it to different water supply regimes, including current average supplies, current extreme-year supplies, and projected future surface water flows under a changing climate.[12] Many of our water supplies are stressed—that is, demands for water outstrip natural supplies in over 9 percent of the 2,103 watersheds examined in this study.

Trends

During the forty-plus years since that first Earth Day, I've reviewed environmental impacts, written natural resource inventories and habitat reports, delineated wetlands, and evaluated watershed conditions—and witnessed the steady erosion of natural resource protection in priority, policy, and practice. Accompanying this decline is a sense of disempowerment among people concerned about local natural resource protection.

Thanks to ongoing scientific research, we have a better idea than ever before about how our activities affect natural systems and the benefits they provide for human communities. However, governments and corporations address issues of human and ecological welfare in the context of political and economic systems that tend to devalue science and the need for environmental protection.

As human communities grow, so do our demands for natural resources. A growing population requires more land, consumes more food, creates more waste, and uses more energy and water. Our current regulatory system for protecting the environment has not been responsive to new information about natural resources or the impact of development activities. It allows corporate interests to sidestep regulations because of promises of economic gain.

Development (whether residential or energy) is not intrinsically negative, but the evaluation of its effect on the environment often falls short of accounting for all impacts. In many cases, environmental-impact evaluations are designed to justify a project and facilitate its passing through the required regulatory hoops as quickly as possible.

Cumulative Impacts and Environmental Regulations

Conservation of natural resources requires looking at long-term effects over large areas, but most land-use decisions are made about small areas and short-term effects, one parcel at a time. Our current regulatory system is geared toward individual actions, while cumulative impacts are piling up to wreak serious havoc on our environment. The ecological footprint is one way of measuring these impacts. Simply defined, this is the area of land and water required to produce all the goods we consume and to assimilate the wastes we produce (including carbon emissions). All the human activities and their impacts that make up the ecological footprint have a cumulative effect. Research by the World Wildlife Fund indicates that by the 1980s our global ecological footprint had reached the capacity of the earth to provide resources and absorb wastes, and by 2003 had overshot that capacity by 25 percent.[13] Americans have among the largest per capita ecological footprints in the world, and scientists have documented extensive ecosystem degradation across the country.[14] Moreover, according to the UN Millennium Assessment, humans are degrading 60 percent of global ecosystem services or using them at levels that can't be sustained. These services include freshwater supply, fisheries, air and water purification, and the regulation of natural hazards and pests.[15]

When we try to evaluate natural resources by cost-benefit analyses, we lose the capacity to examine the health of ecological systems over time. Joseph Guth presents an excellent overview of cumulative impact evaluation and cost-benefit analysis from a legal perspective.[16] Guth indicates a need for a major shift in thinking about how we protect the environment. His points include the following:

- Our current environmental laws are designed to promote limitless economic growth and assume that this growth provides a net benefit to society, despite any damage it may cause (to the environment). This assumption does not take into account the fact that the earth has a limited capacity to absorb environmental degradation. Our laws are not designed to maintain ecological systems or services.
- Cost-benefit analysis doesn't work in favor of environmental protection; one of the reasons for this is that it doesn't take into account cumulative impacts: "The law will have to abandon its use of cost-benefit analysis to justify individual environmental impacts and instead adopt the goal of maintaining the functioning ecological systems that we are so dependent on."[17]
- The law favors environmental and human health protection—when analysts can prove that benefits of protection outweigh costs in terms of dollars. Whenever plaintiffs or regulators can't carry this cost-benefit burden of proof, activities that damage the environment can proceed.

Since their inception, environmental-protection regulations have been subject to political scrutiny. Legislators have subsequently weakened some regulations and repealed or underfunded others to the point where enforcement is impossible. Where adequate regulations and laws do exist, the reality of the political context in which they are implemented dictates how effective they are. As one writer observed, "During 2011, the US House of Representatives voted nearly 200 separate times to block, delay or weaken the commonsense safeguards we all depend on to protect our waters, wildlife, lands and air. It was the single worst legislative assault in history on the foundational protections set in place over the past four decades."[18]

Luckily, many of these initiatives didn't make it beyond the House. But the attitudes that spawned them trickle down to affect state and local regulations and policies. For example, local governments in New York State have become increasingly reluctant to consider adoption of local water resource protection ordinances. These laws attract controversy and are lightning rods for misinformation and local conflict. Where such laws have passed, they are often challenged in court. Their staying power is tenuous, despite immediate evidence of the need for local water protection to address increased flooding and water pollution.

Clean Water Act

In 2012, the federal Clean Water Act turned forty.[19] It was born in an era of landmark environmental protection initiatives in the United States that collectively set out to protect clean air, clean water, and the plants and animals affected by human activities. Since that time, many states and local municipalities have passed their own versions of these acts, seeking further protection for water resources. Citizens support a range of local and regional initiatives that improve water-resource protection, including river keepers, watershed alliances, river walks, and stream biomonitoring. But we still have a long way to go.

At a recent regional Watershed Alliance conference, one of the speakers asked the audience, "How many of you think that the Clean Water Act has been doing a good job of protecting our water resources?" Only a few people (in a room of over a hundred) raised their hands. I'm not going to debate the effectiveness of the Clean Water Act and its amendments here, nor am I going to guess what was going on in the minds of those conference participants. However, as a wetland professional, I do know that at a national level we have thus far failed to protect our wetlands in a way that ensures their future existence. The Clean Water Act does not guarantee water-resource protection. Instead, water protection faces unremitting challenges.

For example, since 1990, US policy under the Clean Water Act has been "no net loss of wetlands." To fill a wetland, you need a permit from the US Army Corps of Engineers, and if the potential impact is serious enough, the proposal requires a detailed study. But the coal-mining industry has often obtained the necessary permits to dump wastes into wetlands and streams without due consideration of possible environmental impacts. How is this possible?

In one case, brought by West Virginia environmental groups against four Massey Energy mining projects, the Army Corps of Engineers conceded that it routinely grants dumping permits with virtually no independent study of the possible ecological impacts and relies instead on the coal companies' environmental assessments. In a 2007 decision in that case, the judge found, "The Corps has failed to take a hard look at the destruction of headwater streams and failed to evaluate their destruction as an adverse impact on aquatic resources in conformity with its own regulations and policies."[20] But because three of the mining projects challenged in that case were already under way, the judge allowed them to continue, pending the case's resolution. Massey appealed the case.[21]

We create environmental-protection regulations to afford some protection, but we also create political and economic systems that influence those regulations and how they are enforced. Many of our elected officials and the media have relegated environmental protection to "special-interest group" status, the concern of an elite group of radical scientists and activists, or disgruntled homeowners who like to complain. But environmental protection is, in reality, a concern for us all. The earth is our home, our habitat, and its condition affects us whether we are aware of it or not, whether we live in a high-rise apartment, a suburban subdivision, or a small rural community.

Economy versus Environment

How often have you heard that you can't have both a healthy economy and a healthy environment? Yet this choice is based on false premises. Protection of a healthy natural environment gives us an "infrastructure" of air, land, and water systems that sustains our lives. The proposition of a choice between "economy or environment" reflects a lack of understanding regarding the necessity of preserving this infrastructure. It ignores the real cost of environmental degradation and pollution. For example:

- Long-term air pollution can reduce earning potential, according to a report from the National Bureau of Economic Research.[22]
- Not investing in the environment can lead to higher bills and taxes for future cleanup of contaminants and lost ecosystem services like flood control.[23]
- The rising demand for "green" chemicals comes from the marketplace as customers demand safer products.[24]
- Planning for smart growth and sustainable communities creates places to live that are healthier, more appealing, and more prosperous.[25]
- Environmental degradation affects public health.[26]

We pay an economic price for failing to understanding the consequences of environmental contamination—human health costs; cleanup of contaminated sites; lower property values; lost opportunities for recreation and tourism; and compromised food safety.

An accompanying trend devalues science and compounds this lack of understanding, blurring the distinction between opinion, political expediency, and scientific fact. This narrow view has a direct impact on environmental protection, affecting the way we evaluate impacts from a wide range of development projects. When "expert" opinions conflict, how do we know whom—or what—to believe?

What Is at Stake?

The consequences of today's environmental-protection decisions usually have long-term effects, shaping what our communities will be like in the future. This example from Harker and Natter's *Where We Live* sums up concern about the environment: "Philip Black of Pendleton County, Kentucky, says he became actively involved in solid waste issues in his community because of concerns about the water in the creek running by the landfill. In his words, 'You know, my grandparents swam in that creek, my parents swam in that creek, my wife and I swam in that creek. If one kid gets sick from swimming in that creek twenty years from now because of something I didn't do, that would be terrible.'"[27]

Black's simple statement of concern underscores the reality that we are all connected, and that as individuals, each of us has a role to play in protecting our water. We all depend on a healthy environment for our physical, mental, and spiritual well-being. This dependence may be as obvious as the need for clean drinking water, or it may be more subtle. Richard Louv has compiled extensive research indicating that direct exposure to nature is essential for healthy childhood development and for our physical and emotional health. In *Last Child in the Woods*, Louv links the absence of nature in the lives of children to disturbing childhood trends, including the rise in obesity, attention-deficit disorders, and depression. This link between health and nature isn't limited to children.

Spurred by worldwide concern about the condition of the earth's ecosystems, the United Nations began a project in 2000 called the Millennium Ecosystem Assessment. It involved over thirteen hundred experts from ninety-five countries and in four years produced a report, *Ecosystems and Human Well-Being*, that represents an up-to-date, comprehensive review of the earth's ecosystems and their condition. The report discusses how human activities change ecosystems and the consequences of those changes for human well-being, and it also establishes a scientific basis for conservation. Like the settlers who saw the American West as an endless frontier of land and space, we may want to believe the natural systems that sustain us will continue forever, beyond the reach of impacts from our activities on the land. Unfortunately, that is not the case. The UN report presents this reality in definite terms:

> At the heart of this assessment is a stark warning. Human activity is putting such strain on the natural functions of Earth that the ability of the planet's ecosystems to sustain future generations can no longer be taken for granted.

As human demands increase in coming decades, these systems will face even greater pressures—and the risk of further weakening the natural infrastructure on which all societies depend.

Protecting and improving our future well-being requires wiser and less destructive use of natural assets. This in turn involves major changes in the way we make and implement decisions. . . .

Above all, protection of these assets can no longer be seen as an optional extra, to be considered once more pressing concerns such as wealth creation or national security have been dealt with.[28]

Our challenge is to bridge the gaps between the information generated by those who study complex, interconnected natural systems and those who make the decisions that will affect those systems.

Why Do We Need a Grassroots Solution?

Protection of air, water, and ecosystems requires the involvement and empowerment of people at the local level. Many people I work with recognize a need for environmental protection but don't know what to do, or feel powerless to make a difference. If energy development, jobs, and promises of economic relief give short shrift to environmental protection, how do citizens level the playing field? As we strive to protect the environment that sustains us, it is important that we "work smart"—maximizing the use of our time and our chances for success.

Many people are aware of this. A national survey by the Pew Research Center for the People and the Press, conducted among fifteen hundred adults in February 2012, found that 50 percent of those surveyed wanted the federal government to strengthen environmental protection regulations, and 29 percent wanted regulations kept as they are. Only 17 percent were in favor of reducing these regulations. A Nature Conservancy poll, conducted in July 2012 by two opinion-research firms (one Republican, one Democratic), found the following:

- Seventy-nine percent of poll respondents say the United States can protect nature and have a strong economy at the same time.
- While 80 percent of voters say the economy is a serious problem, 74 percent don't want to cut federal funding for conservation.
- Voters are more likely to say that protections for land, air, water, and wildlife have a positive impact on jobs (41 percent), than a negative impact (17 percent), or little impact one way or the other (33 percent).[29]

Often we lose sight of the connections between land-use activities, ecosystems, and the benefits natural systems provide. If you were to ask a group of community residents "Who is in favor of protecting an adequate supply of clean drinking water?" all of them would probably raise a hand. But if you present that same group with the steps that they need to take to actually protect the community's water, and then ask "Who is willing to support these actions?" few are likely to raise a hand in affirmation.

Many people don't believe that water-resource protection is an urgent problem. Plenty of people in the United States have unlimited water at the touch of the tap. But even where water is abundant, we have no guarantee that it will remain so into the future, unless we take care of it now. And just because we have an adequate water supply doesn't mean it's free of contaminants.

When we look at natural resources in this way, their protection is no longer a partisan or elitist or scientific issue: It is a personal issue for each of us. What we buy, how we vote, where we recreate—all these decisions affect the natural world. And in turn the natural world affects you and me.

As long as most of us can still see beautiful landscapes, recreate in protected parks, or enjoy a glass of clear water, we want to believe that all is well with the natural environment. No one welcomes bad news. And optimism is better energy than pessimism. But despite our desire to believe that all is well, the world I work in as a natural resource consultant shows a different side: wetlands filled in, streams polluted, little or no regulatory oversight, fear of lawsuits driving local planning decisions. None of these trends will be reversed without significant public pressure.

A *New York Times* article in July 2011 discusses the acidification of the oceans and its implications for human communities.[30] The writer describes the protection of the oceans as "the work of nations, but such goals require pressure from ordinary citizens if there is to be any hope of bringing them about in the face of opposing political and economic interests." These "opposing political and economic interests" permeate the information we receive about all types of development projects, generating confusion about impacts.

We all have a right to an adequate supply of clean water. We also have a right to enjoy the many benefits provided by natural water systems, but these systems are not automatically protected by existing regulations. Decisions about development—residential, commercial, energy—can dramatically affect the place where you live, including your supply of water. Residents have the power to affect those decisions. People in communities across the country are beginning to realize that their actions can make a difference. More people are getting involved in local action to protect the environment as it becomes clear that government agencies can't provide sufficient protection and that the existence of regulations does not ensure effective protection. People are also becoming aware that government agencies as well as corporate and business entities don't always operate in the public's best interests—even when it comes to protecting our health via the food we eat, the air we breathe, and the water we drink.

Connecting the Drops

The business of protecting water and wet places is based on an understanding of interconnections. It requires both a broad view and attention to details. It includes consideration of different perspectives—from the top of a mountain for a look at the streams and lakes below, to a trek through the foot-squelching mud of a marsh for a closer look at the creatures disappearing into the water weeds. Both of these views are necessary to understand the extent to which we have collectively messed up the natural systems we depend on, and what it will take to protect what remains.

We can no longer leave protection of our water resources solely to government regulations, expert consultants, or environmental-impact reviewers. The problem has become too big for that. Water is a shared resource. Because we will always need it, we will always need to protect it.

This work challenges us to

- become familiar with basic information about ecosystems and watersheds, how they work, and the benefits they provide for human communities;
- describe specific threats or impacts to those systems (for example, from residential and energy development activities);
- develop and implement strategies for action to protect natural resources;
- interpret and understand conflicting information about environmental issues or development projects;
- present an informed, organized case to decision makers, based on scientific information;
- initiate consideration of cumulative impacts and true costs during environmental-impact review;
- develop a position of strength based on solid information—rather than on the politics that insist on making the protection of clean air and water a partisan issue; and
- counter claims of "it's a choice between the economy or the environment."

Connecting the Drops presents the basics of water resource protection: ecology and watershed science; techniques for evaluating environmental impacts; obstacles to protection and how to overcome them; and tips for protection strategies that maximize chances for success. Throughout this book, I share my experiences as a natural-resources professional, a landowner, and a volunteer. I share examples of actual projects and local actions from my over twenty-five years of experience with environmental-impact assessment, wetland research and delineation, water resource protection, and biodiversity assessment. Since what we do on the land affects water, I do not limit my approach to aquatic or wetland ecosystems; I also consider upland systems and adjacent land use. The methods described in *Connecting the Drops* have broad application in diverse geographic locations. The environmental details may differ, but the methods are the same. I use residential development and hydrofracking for natural gas as examples because they include a range of impacts shared by other types of development.

The chapters are arranged in an environmental-protection progression that begins with describing natural systems and their benefits. After establishing this foundation, we consider specific development activities and how they may affect these natural systems and benefits (that is, how they can affect you). Then we evaluate this information so it can be used for developing a strategy or plan and answering questions such as

- Will impacts be significant?
- Will mitigation be effective?
- Will true costs of development be considered?

This book is divided into three sections as follows. Part 1 provides the foundation—information about natural water systems.

- Chapter 1: Ecosystems, watersheds, and natural cycles, how they work, and the services and benefits these systems provide for us
- Chapter 2: How to identify and describe natural systems on a particular site, focusing on soils, watersheds, and water-based ecosystems (wetlands and streams)

Part 2 leads you through the process of how to describe and identify impacts on the natural systems described in part 1.

- Chapter 3: How to describe impacts of specific development activities on watersheds and ecosystems, focusing on residential development

activities and their effect on water resources before, during, and after construction
- Chapter 4: An in-depth look at four major water-protection issues—water contamination, flooding, groundwater loss, and decrease in biodiversity—and the connections between these issues and land-use activities. (The appendix provides sample checklists for describing impacts on a wetland and a small stream.)
- Chapter 5: How to describe impacts of energy development on water resources and systems, with the focus on sorting out the complex array of activities associated with hydrofracking for natural gas. The chapter also includes examples from mountaintop removal of coal, oil pipelines, wind farms, and hydroelectric projects.

Part 3 focuses on how to evaluate impacts, interpret and share information, and translate the information discussed in previous chapters into action.

- Chapter 6: Significant impacts, cumulative effects, and mitigation—what these catchwords mean for water resource protection, and how to develop criteria to determine when environmental damage is significant.
- Chapter 7: Obstacles to effective protection—and how to overcome them. These include inadequate information, controversy and conflict, lack of political will, and property rights issues.
- Chapter 8: How to develop a strategy for protecting resources—bringing people together, maintaining credibility, interpreting information, and communicating effectively.

As you progress through this book, remember that you are not alone in your concern and frustration or your desire to make a difference. In the words of Amitai Etzioni in *The Spirit of Community*: "We have a moral commitment to leave for future generations a livable environment, even perhaps a better one than the one we inherited, certainly not one that has been further depleted."[31]

part I

HEADWATERS

Understanding
Natural Water
Systems

1

Natural Water Systems
and Their Benefits

The environment is not a competing interest to be balanced with
other interests; rather, it is the playing field, the very foundation,
upon which all our interests compete.

DR. MICHAEL KLEMENS, 2001

Natural resource systems—ecosystems, watersheds, cycles—affect the
water we use every day. By understanding the basics about these systems and how they work, we can improve the process of matching land-use
activities with their specific effects on natural systems and the benefits these
systems provide. Depending on your situation, this chapter may provide all
you need to know, or it may guide you to sources of additional information
(see the appendix).

Recent events in a small New York town illustrate the need for this overview of natural systems. Town leaders drafted a local water resource protection law and invited the public to respond. But instead of commenting on the
procedure for protecting the town's water, residents were asking more basic
questions: "Why do we need to protect buffers around wetlands and streams?
They're just one more constraint on land development." "This law protects
wetlands of all sizes, so that means any puddle can be a wetland." "We have
plenty of water, so what's the problem?" "Nature recovers from disturbance
on its own."

These are common questions and comments. While most people will agree
that they want to protect their water, a lack of basic information leads to uninformed reactions and decisions. It can also lead to further degradation and
contamination of water resources, because the tendency is to either refuse to
adopt local protection regulations or to weaken them to the point where they
no longer provide actual protection.

In this case, residents were unprepared to discuss the finer points of the
proposed new regulations; they didn't understand the underlying science that
clarified why particular protection actions were necessary. This situation

forced the group proposing the regulations to backtrack from soliciting comments to filling in key information gaps.

To help you avoid a similar scenario, this chapter will provide a foundation for understanding natural systems, their interconnections, and the services they provide.

The Water Cycle

More than half the human body is composed of water. That water carries nutrients, removes toxins, and keeps us healthy and functioning, right down to the cellular level. But our bodies are not test tubes, and the water doesn't stay contained. We're part of a much larger system—the water cycle, which is key to maintaining the environment outside our bodies. The water in our bodies connects us to all other water—lakes, ponds, aquifers, rivers, streams, and wetlands that carry and store water in the places where we live. The water that fills the glass in your hand—where did it come from, and where is it going?

More than 97 percent of the earth's water is salt water; about 2 percent is tied up in glaciers and icecaps, and almost all the remaining water is stored as

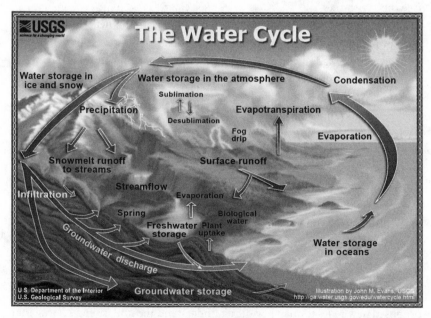

FIGURE 1.1. The water cycle. Courtesy of the US Geological Survey: Department of the Interior, USGS.

groundwater. Only a small proportion (about 0.01 percent) of the planet's water is in surface waters such as rivers, wetlands, and lakes.[1] Water is constantly on the move, falling on the land as rain or snow, seeping into the ground, or flowing across the surface of land or pavement until it reaches a stream or pond. Water also cycles through plants, animals, and microorganisms; returns to the air through evaporation or transpiration (the passage of water through a plant from the roots through stem and leaves, and into the atmosphere); and filters through the soil. Natural processes, along with other cycles in nature (chemical, physical, biological), clean the water and replenish the supply available for ecosystems and human communities. Water eventually ends up in the ocean, which plays a huge role in the water cycle. Although the focus of this book is freshwater, keep in mind that contaminated rivers, drought, and floods ultimately affect the condition of the oceans as well.

Human activities alter this cycle—what we do on land affects water. We can change land features (remove vegetation, fill in a pond, divert a stream) or change the water's path, sending it through septic systems, wastewater-treatment plants, industrial processes. You are part of this cycle, whenever you drink a glass of water, take a shower, water your garden, flush the toilet, wash a load of laundry, or buy a product that required water for its production (from food and clothing to toys and electronics). Each of us uses an estimated eighty to one hundred gallons of water per day. If your house sits on a half-acre lot and a storm drops one inch of rain, that's about 13,577 gallons of water in your yard, if you collect all of it.[2]

Even after we use it, water returns to the cycle. When we contaminate water or change the way it moves through natural systems for purification and replenishment, we reduce the natural systems' ability to work well (in this case, purify the water), and we suffer the consequences of water pollution. You can see the effects in very small waters, like the stream in your backyard, or in very large areas, like Chesapeake Bay, the Colorado River, and the Gulf of Mexico.

Natural Systems

When we describe natural resources in terms of "systems," we are focusing on how nature works, not just on how it looks. By understanding natural systems that haven't yet been changed, we have a baseline of conditions against which we can compare impacts of land-use activities: industrial soaps from a car wash; erosion from a new house lot; nitrogen, phosphorous, or pesticides from a garden or farm. Altering a natural system will likely affect its benefits as well.

For example, a fisherman catches a trout in a stream. Trout can survive in streams only if stream flow is adequate and the water is clean and cool, with high levels of dissolved oxygen. Adjacent land use, air pollution, and the extent of nearby human activity affect a stream's water quality. By understanding basic connections between activities that cause pollution and the quality of water required by trout, we are better able to persuade others to value the stream that supports the fish by protecting the land along its banks—and even addressing air contaminants that can pollute water. You may want to eat that trout, but if mercury contaminates its stream and shows up in the trout's flesh, the trout's contamination becomes your personal health issue. Sources of mercury contamination include coal-fired power plants, boilers, cement plants, and other sources that spew it into the air. From the air, mercury falls into the water, where it enters the food chain and eventually the fish. Mercury affects not only the fish's body but can poison your body as well.

Mercury is only one among many water-quality contaminants that have the potential to affect natural systems and human health. Why investigate contaminants yourself? Why not just hire an expert to produce a report for you? In fact, you may need to do both. Specific local situations may require a professional report or an in-depth study. However, you can save time and money by having a clear idea of which issues you need to pursue, what kind of expert to hire, or even what questions to ask about water resources and impacts on them. And if an expert is not immediately available, you can start the process anyway.

Ecosystems

I would like to use a simpler term—but there isn't one. Unlike the more general terms "nature" or "environment," the scientific term "ecosystem" refers specifically to a working system of interrelated parts, both living (plants and animals) and nonliving (soil, nutrients, air, and water). An ecosystem is a community of animals and plants interacting with one another and with their physical and chemical environment. Habitat is the place (within an ecosystem) where a plant or animal lives. Physical and chemical conditions allow particular plants to grow; plants in turn provide a variety of habitats for animals. Ecosystems may be very large (for instance, open ocean, tropical rain forest) or small (pond, wetland, or stream). Living organisms in an ecosystem range from large (trees, eagles, lake trout) to microscopic (algae, bacteria).

An ecosystem is a busy place. You can easily see its obvious features, but the parts you can't see power the entire system and make it work. As figure 1.2 shows, plants transform energy from the sun into food; animals consume

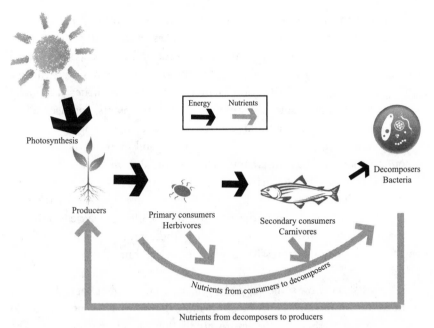

FIGURE 1.2. Diagram of a simple ecosystem. Illustration by Angela Sisson.

plants and other animals; and nutrients cycle through plants, animals, air, water, and soil or sediment.

Here's a glimpse of what is going on among the three basic groups of organisms in an ecosystem:

- *Producers*: Algae and plants convert energy from the sun into living plant matter by the process of photosynthesis. Within these producers, light energy (from the sun) and chlorophyll (from the plant) transform water and carbon dioxide (from the air) into food and oxygen, which is released into the air.
- *Primary and Secondary Consumers*: Animals eat plants or other animals. These consumers convert food and oxygen (from the air) into energy and carbon dioxide (released into the air).
- *Decomposers*: Bacteria, fungi, and small invertebrates break down dead plants, animals, and wastes and return organic nutrients to the ecosystem for plants and animals to reuse.

As these organisms interact with one another, they also play a role in water, carbon dioxide–oxygen, and nutrient cycles that sustain ecosystems over time. For example, plants and animals need nitrogen to build protein. The chemical element nitrogen, as nitrogen gas, makes up about 80 percent of our air, but plants can't use it in that form. Nitrogen-fixing bacteria or algae change that nitrogen from the air into ammonia. Other microorganisms, through the process of nitrification, change ammonia into nitrites and finally nitrates, a form that plants, and animals that eat the plants, can use. When plants and animals die or animals excrete waste, decomposer organisms process the waste and release nitrogen back to the soil or water as ammonia, where microorganisms transform it back into useful form as nitrates, so that it can cycle again through plants and animals. Another set of microorganisms, through the process of denitrification, transform nitrates back into nitrogen gas, which returns to the atmosphere. The nitrogen cycle is only one of many ongoing processes in ecosystems. These processes sustain life on earth and support an impressive diversity of living things, including human beings.

Table 1.1 shows how some basic physical, chemical, and biological features of an ecosystem interact. Ecosystems are more than a "cast list" of animal and plant characters. They are the entire production—including the script, the role each character plays, and their relationships in the overall plot. Within ecosystems, each plant and animal species that lives in communities has a role. Community members are interrelated—so when you change one, you also change its interconnections with other community members and with its habitat.

TABLE 1.1. Examples of stream ecosystem components and interactions

Ecosystem components	Examples	Interactions
Physical	Water depth, current, bank slope and stability, substrate (e.g., sand, gravel, silt)	Flowing water is habitat for aquatic plants and animals; undercut banks provide shelter for fish; rocks on the bottom shelter aquatic insects.
Chemical	Dissolved oxygen, pH, suspended sediment, dissolved nutrients	Changes in water chemistry affect the ability of aquatic plants and animals to thrive in the stream.
Biological	Riparian shrubs and trees, aquatic plants, algae, fish, aquatic insects	Trees stabilize banks; plants and animals recycle nutrients; animal waste affects water chemistry.

Changes to these interconnections can have unexpected consequences. For example, in a 2005 study, when wolves were reintroduced to Yellowstone National Park, the nesting success of songbirds along rivers increased. Where wolves are scarce or absent, deer and elk populations increase and overgraze lush trees and shrubs along streams. This grazing damages the riparian vegetation that is critical nesting habitat for migratory songbirds.[3] Along stream edges, the loss of trees and shrubs also removes shade that keeps the water temperature cool enough for cold-water fish like trout. Trees and shrubs stabilize stream banks and reduce erosion, especially during seasons of high water.

Biodiversity

Healthy ecosystems support a diverse array of plants and animals. For most of us, this biodiversity simply means the variety of living things in a particular area, beginning with plants and animals we can easily see. But biodiversity also includes species that we don't typically see—those that live underground or underwater, wildlife that comes out only at night, microscopic organisms, and rare species. Technically, biodiversity also includes genetic diversity, seen among individuals within a species (for example, every human being has a different face), and ecosystem diversity, seen in the diversity of habitats in a particular region (the more habitat types, the more species). Biodiversity conservation often focuses on protecting very large areas; but if we are to protect species richness we also need to protect diversity in smaller habitats within our communities. Connected habitats like small streams that weave through suburban neighborhoods can support a significant variety of plants and animals. Protection of a region's biodiversity often depends on how well we preserve habitat in our communities.

Roles

Plant and animal diversity stabilizes an ecosystem and keeps it working. These plants and animals are like the actors in theater, maintaining the action and intrigue of a play. If one or two characters temporarily step out, others may step in to take up their roles and keep the story moving forward. But if a cast of twenty drops to five, there are no longer enough characters to play all the roles, and the story stumbles. Similarly, loss of species in an ecosystem disrupts the system's functions and equilibrium. This lack of balance in turn affects the benefits the system provides for human communities.

The roles that different species play in their community or ecosystem keep the system functioning. We can provide a list of species found on a particular site, but this is of limited value unless we also look at how they fit together and interact within the ecosystem. Rare species as well as common ones are important to the system. Keeping common species common—and off the threatened and endangered lists of the future—is an important goal for maintaining biodiversity and healthy ecosystems. Some species play leading roles in ecosystems; without them, the system can no longer work in a healthy manner. These "keystone" species affect all other species and even their habitats; loss of a keystone species will change the rest of the ecosystem. Often but not always, keystone species are predators.

Predators and prey exist in a balance: Predators require prey for food and keep the prey species population in check. When predators leave the system, this balance is upset, and prey populations may increase to the point where they damage other parts of the system. Predators are especially important in maintaining how an ecosystem functions; when they are absent, the ecosystem dramatically changes. In the Northeast, for example, the lack of wolves or mountain lions as natural predators to control white-tailed deer populations is one reason for the proliferation of deer. In many places deer browse understory vegetation so thoroughly that they destroy habitat for nesting birds.

The more we understand about roles and connections in nature, the better we can identify how environmental impacts affect them. Since the winter of 2007, millions of insect-eating bats in the United States and Canada have died from a disease called white-nose syndrome. You may not be too concerned about a sudden decrease in the number of bats flying around your neighborhood at night, unless you understand how much they help us. A bat can eat up to its body weight in insects daily, and many of these insects are harmful to crops. Each night during active season, a single little brown bat can consume four to eight grams of insects. This translates into millions of dollars in insect-removal services for US farmers.[4]

Species that are very sensitive to habitat changes and pollution indicate contaminants in the environment. Their presence or absence can alert us to specific environmental conditions, and like the proverbial canary in the coal mine, they provide information about the quality of human habitat. For example, the absence of certain aquatic insects, such as mayflies, may indicate poor water quality in a stream or siltation of bottom habitat. Amphibians and their eggs are sensitive to waterborne pollutants and can be important indicators of water quality.

Native versus Nonnative

A species is native to the area where it evolved. Native plants and animals have evolved together, interacting and shaping one another's development. They compose a system of environmental checks and balances that keep an ecosystem stable. This balance may change when a plant or animal is removed from its original community and moved to a place with plants, animals, and diseases with which it has never interacted. The ecological connections that kept it in check and made it a contributing member of an ecosystem are no longer present. A nonnative population may spiral out of control because of a lack of natural predators.

Nonnative species can dramatically affect native species and their habitats, becoming "invasive." A Cornell University study estimates that about 42 percent of our threatened and endangered species are at risk because of displacement by, competition with, and predation by invasives. The same study reports that damage from some fifty thousand invasive species costs the United States at least $120 billion a year in economic losses (environmental damage and cost of control).[5] Regional studies detail these costs. In the Great Lakes, damages attributed to invasive species cost at least $138 million annually; the figure may be as high as $800 million.[6] Since 2006, the US Fish and Wildlife Service and its partners spent more than $6 million addressing the problem of pythons and other large constrictor snakes in Florida's ecosystems.[7] Eurasian watermilfoil, an aquatic invasive plant, reduced lakefront property value in Vermont by up to 16 percent, and in Wisconsin by 13 percent, according to a 2012 US Fish and Wildlife Service report.[8]

Ecosystem Services and Benefits

Healthy ecosystems can sustain themselves over time, continuing to provide services and benefits. People and their communities affect, and are affected by, ecosystems; ecosystem health affects human health. We depend on ecosystems because of the services they provide, outlined in the following list of categories developed by the Millennium Ecosystem Assessment; the United Nations initiated this project to assess the consequences of ecosystem change for human well-being and to establish the scientific basis for action needed to conserve these systems.[9] The assessment involved more than 1,360 experts worldwide. Their findings provide a state-of-the-art scientific appraisal of the condition and trends in the world's ecosystems and the services they provide.

Supporting Services

Some ecosystem services—soil formation, photosynthesis, nutrient and water cycling—are so basic and universal that they make life on earth possible. They are necessary for the production of all other ecosystem services:

- *Soil formation.* The National Cooperative Soil Survey identifies and maps more than twenty thousand different kinds of soil in the United States. These soils differ from one another depending on how they were formed. Once soil is formed, ecosystem processes and nutrient cycling maintain the soil's fertility. Soil affects where plants grow and how well they thrive and, in turn, is the basis for many habitats and for animals that use plants for food, shelter, breeding sites, and resting areas.
- *Primary productivity and photosynthesis.* To create and accumulate new organic matter, through photosynthesis, plants and microorganisms convert sunlight into energy and store it in plant tissue. Photosynthesis is a key process for ecosystem functioning. The productivity of an ecosystem is the rate at which plants and microorganisms create organic matter, measured in kilocalories (kcal / square meter / year). In general terms, the greater the plant diversity, the more productive the ecosystem. Estuaries, forested swamps, and marshes are among the most productive ecosystems.
- *Nutrient cycling.* Ecosystems sustain cycles between living things and the nonliving components of their environment. The earth's biogeochemical cycles that are critical for life include water, carbon, nitrogen, phosphorus, and oxygen. Nutrients cycle through soil, air, and water into plants and animals, and then back into the physical environment. Living organisms change nutrients as they pass through these cycles.
- *Water cycling.* Water, essential for all living organisms, continuously cycles through ecosystems. Through this cycle, water maintains all living things, is purified, replenishes surface and groundwater supply, and transports nutrients.

Provisioning Services

This group of services yields the products we obtain from ecosystems:

- *Food.* Ecosystems sustain plants and animals that we use for food.
- *Pharmaceuticals.* Many species of plants, animals, and microbes produce or contain chemicals that are used in pharmaceuticals, botanical medicines, and biological pest control.

- *Biomedicine.* Medicines derived from natural ecosystem biological and biochemical processes are important to human health.
- *Freshwater.* Like plants and animals, humans require clean and abundant water.

Regulating Services

These ecosystem services regulate and balance ecosystem processes:

- *Disease control.* Changes in ecosystem processes—for example, a decrease in plant and animal diversity, or changes in surface water storage—can increase organisms that carry and transmit human disease. A system with a greater native species variety can better control undesirable species. For example:

  West Nile virus. The northern house mosquito transmits West Nile virus; the mosquito breeds in standing water collected in rain barrels, clogged gutters, flowerpots, discarded tires, catch basins, and other containers where predators are absent. In contrast, in healthy wetlands a diversity of predators can keep mosquito populations in check. For example, when the Essex County Mosquito Control project in Massachusetts restored a fifteen-hundred-acre wetland, the mosquito population dropped significantly.[10] In some cases, water management practices to increase mosquito predators have reduced the cost of mosquito control by more than 97 percent over the traditional method of insecticide application.[11]

  Lyme disease. Small woodlots contained an average of three times more deer ticks than larger forested patches and seven times more ticks carrying Lyme disease, according to the Cary Institute in Dutchess County, New York. Small woodlots in developed areas support lower biological diversity—fewer predators and less competition from other small mammals leads to a proliferation of white-footed mice. These mice are the prime hosts of bacteria that cause Lyme disease. Ticks pick up the bacteria from the mice and transmit them to people.[12]

- *Biological pest control.* Natural systems maintain an insect predator-prey balance for biological control of agricultural crop and livestock pests and diseases. Natural insect predators can often control pest species, such as insects that damage agricultural crops, without using pesticides that cause other ecological damage.

- *Pollination.* Flowering plants (including about 75 percent of crops) depend on bees, bats, hummingbirds, butterflies, and other pollinators, which in turn thrive when the ecosystem meets their habitat requirements. Pollination services have high economic value; for example, insect-pollinated crops contributed at least $20 billion to the US economy in 2000.[13] Pollinators keep plant communities healthy, contributing to habitat value, erosion prevention, and water quality.
- *Protection from natural hazards.* Healthy ecosystems moderate the effects of natural hazards such as floods, storm surges, and fires. For example, wetlands reduce floods by absorbing peak runoff, and barrier beaches and dune systems buffer wave impacts from storm surges.
- *Water purification and waste treatment.* Ecosystems filter out and decompose organic waste and take up or detoxify chemicals through plant and soil processes.
- *Water regulation.* Healthy watersheds regulate flooding, aquifer recharge, stormwater runoff, and water storage.
- *Erosion.* Plants retain and stabilize soil and prevent landslides.
- *Climate regulation.* Changes in plant cover can affect local temperature and precipitation. Soil and plants store carbon from the atmosphere.

Cultural Services

Ecosystems also provide nonphysical benefits for human communities, including the following:

- *Quality of life.* Spiritual and aesthetic experiences contribute to well-being; local "rural character" may also contribute to increased property values.
- *Education and research.* As we continue to learn about the natural world, and how human activities change ecosystems, intact natural systems allow us to discover new information about their benefits and how to protect them.
- *Recreation and nature-based tourism.* Natural areas attract us for vacations and a host of outdoor activities. Ecosystems sustain the natural resources that invite visitors and support fishing, hunting, bird watching, photography, and hiking.

Biomimicry

"Innovation inspired by nature," or biomimicry, is another ecosystem benefit. By studying how nature works, we can imitate some of these "workings" and use them as models to solve human problems.[14] For example, the bumps at

TABLE 1.2. Biomimicry examples

Plants and animals	What they do	How we are using it
Water strider, mosquitoes	Strider feet and mosquito eyes have surfaces that repel water	Deicing without using chemical compounds (hydrophobic deicing)
Amazon electric eel, electric catfish	Produces chemicals that generate electric current	Battery-cell designs that mimic the groups of cells that give electric eels their ability to shock; these electrocytes have inspired the design for a product that could power biomedical implants
Kingfisher	Dives from air into water with very little splash to catch fish	Implementing design changes for noise reduction and increased energy efficiency in very high-speed trains
Dragonflies	Wings take advantage of wind for movement and solar energy for warming muscles for flight; wings with dual motion for increased maneuverability	Installing solar sails on boats that switch between wind and solar energy as the power source depending on availability; reduces fuel consumption and emissions

the front edge of a whale fin greatly increase its efficiency, reducing drag and increasing lift. Wind turbine blades based on this design greatly boost the amount of energy created per turbine. Additional applications include more efficient cooling fans, airplane wings, and propellers.

Many biomimicry innovations are related to energy efficiency and storage. Energy technology explores capturing solar energy the way a leaf does and designing sun-tracking solar panels that are inspired by plants and move without motors. Energy applications are described on the New York State Energy Research and Development Authority (NYSERDA) website—"Living systems have optimized the capture, storage, transformation, and use of energy. In an era of rising energy costs and concerns over use of fossil fuels, it's worth looking to nature for how to harvest renewable energy and increase energy efficiency."[15] Between 2005 and 2008, according to NYSERDA, sales of biomimicry-based products and architectural projects generated over $1.5 billion in revenues, and numerous commercial products inspired by nature have also been released. The NYSERDA and Biomimicry Institute websites describe biomimicry projects; examples are shown in table 1.2.[16]

Economic Value

We can measure some ecosystem services and benefits more easily than others (for example, flood protection, primary production, water-quality improvement, and fisheries). By assigning dollars to ecosystem services, we may find that the value of managing healthy, sustainable ecosystems is higher (in terms of economic and public health costs) than the value of converting them to other land uses. Here are a few examples:

Flood damages may potentially cost millions of dollars. In studying Minnesota wetlands and floodplains, the Wetlands Initiative noted that restoring the natural hydrology of floodplains would decrease flood peaks and the cost of damage. Replacing the flood-control function of the five thousand acres of wetlands drained each year in Minnesota alone costs an estimated $1.5 million. Preserving wetlands in the first place and restoring some of those that have been drained could help reduce future flood losses.[17]

The US Army Corps of Engineers uses wetlands to prevent flood damage. In Massachusetts, the Corps calculated that the loss of all wetlands in the Charles River watershed would cause an average annual flood-damage cost of $17 million. It concluded that conserving wetlands was a natural solution to controlling flooding and less expensive than the construction of dikes and dams alone. Therefore, the Corps acquired 8,103 acres of wetlands in the Charles River basin for flood protection.[18]

The Mississippi Delta's ecological communities currently generate up to $13,000 per acre in ecosystem services each year, according to Mississippi Delta Research. These services include hurricane storm protection, water supply, climate stability, food, furs, habitat, waste treatment, and other benefits worth at least $12 billion to $47 billion a year. Wetland ecosystem services account for more than 90 percent of this value.[19]

Watersheds

Whether you get your water from an individual well or a municipal system, all the water you drink comes from a watershed—the area of land that drains into a particular stream, river, lake, or other body of water (figure 1.3). Watersheds contain both land- and water-based ecosystems—forests, fields, wetlands, streams, and lakes—as well as groundwater resources. The health of the watershed as a whole depends on the health of all these systems.

Surface water and groundwater connections and interactions affect the physical, chemical, and biological processes in lakes, streams, and wetlands. Geology and topographic features shape the path of streams and affect water flow and chemistry. Soil conditions determine how fast water flows over the land and how quickly it seeps into the ground to replenish groundwater supplies. Vegetation reduces stormwater runoff and erosion, checks flooding, improves soil and water quality, provides habitat, and in

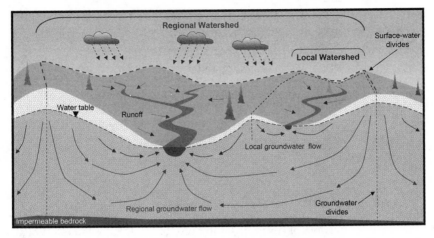

FIGURE 1.3. Watershed and the connections between surface and groundwater. Courtesy of Kevin Masarik, University of Wisconsin–Stevens Point.

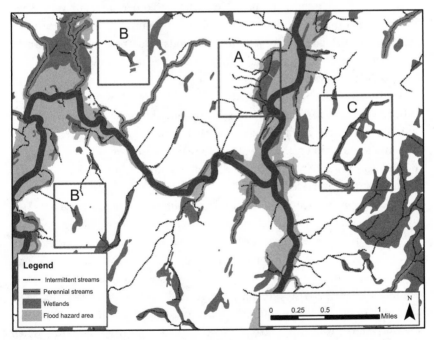

MAP 1.1. Surface water resource interconnections. (A) Several intermittent streams drain into a large wetland in the floodplain and merge into a tributary to the river. (B) Headwaters of these small streams are in wetlands. (C) Intermittent stream system is part of an extensive wetland, connecting it to a large wetland complex and a perennial tributary to the river. Map by Angela Sisson.

some places maintains cool water temperatures. The water features within a watershed affect the location and volume of water stored, rate and volume of stream flow and runoff, and water chemistry.

Map 1.1 shows connections among wetlands, floodplains, and streams. We can trace problems with flooding, drinking-water supply, and pollution to the condition of a watershed's features and their roles in the movement and storage of water. To protect our water, we need to understand how watershed systems work and how all their features are interconnected. The following is a brief overview.

Rivers and Streams

As part of the network of connected water resources, streams and rivers of all sizes provide valuable watershed services. A network of small streams distributes floodwaters from heavy rainfall across the landscape and channels some of it to larger streams and rivers, as well as to lakes, ponds, and wetlands. Nutrients carried in soil, plant parts, leaf litter, and other materials are washed into streams from the land. Transported downstream, they support fish and other aquatic life. During hot weather or periods of low flow in river channels, small tributary streams, especially those with cool and relatively clean water, may provide refuges for fish and other aquatic life.

Headwaters—often associated with wetlands, seeps, or springs—are the initial source of water for all river systems. Small headwater streams collect floodwater or runoff, recharge groundwater, support a high diversity of species, and sustain downstream waters. They comprise just over 50 percent of total stream miles in the continental United States and provide the foundation for all our large river systems.[20]

Water supply in streams comes from precipitation, groundwater, or surface water (smaller tributary streams, seeps or springs, lakes, and wetlands) in varying proportions at different times of the year. Stream flow is measured in cubic feet per second; it normally fluctuates with seasons and rainfall patterns, depending on the location, type of stream, weather patterns, and local conditions. Streams also interact with groundwater. When the water table next to a stream is higher than the surface of the water in the stream, the stream receives water from groundwater (called a "gaining" stream). When the water table is lower than the stream water surface, the stream loses water to the groundwater system (called a "losing" stream). Many streams are both gaining and losing, at different times of the year, depending on precipitation, depth to water table, and other conditions. Groundwater and wetlands maintain downstream water supply and minimum stream flow, which is critical for the survival of aquatic life.

Three types of small streams, described here according to their water-flow characteristics, may change temporarily during periods of extreme flooding or drought:

- *Perennial.* Water flows year-round. Water source is from smaller upstream waters, tributaries, or groundwater; runoff from rainfall or other precipitation is supplemental.
- *Intermittent.* Water flows some of the time, depending on seasonal rainfall. Both groundwater and surface water sources feed these streams.
- *Ephemeral.* Water flows only in response to storms; in arid regions these streams are known as dry washes or swales; precipitation is the only source of water.

While water flow defines aquatic habitats, vegetated areas along streams and rivers serve as transitions between aquatic and upland systems. These riparian zones provide important habitat and support a high diversity of plants and animals. Vegetation (especially trees and shrubs) stabilizes stream banks, provides nutrients for downstream areas, moderates flood flows, slows runoff from surrounding lands, and prevents erosion. Overhanging bank vegetation maintains cooler water temperature, especially in small streams.

Low-lying riparian areas may overlap with floodplains. When a stream overflows its banks, floodplains slow floodwaters and spread them across the land. Standing water soaks into the ground and deposits its load of sediment in the floodplain. The stream may carry nutrients from the land into the system along with receding floodwaters. Floodplains are part of the normal pattern of stream dynamics. Channels, meanders, and flow patterns change over time, and streams shift within their floodplains. Meandering stream channels tend to slow stream-flow; straight channels are more conducive to high-velocity water flows.

Lakes and Ponds

Water that fills lakes and ponds comes from inflowing streams, precipitation, overland runoff, or groundwater. Within a lake or pond, water volume and temperature may vary with seasonal changes and local conditions. Sediment that settles to the bottom can decrease lake storage capacity. A lake's storage volume and water-residence time depend on water gain from precipitation and tributary streams, and water loss through outlet streams and evaporation. In deeper lakes, seasonal water circulation patterns mix deep water and surface water, maintaining water quality and oxygen supply, nutrient cycling, and aquatic life in the lake.

The shallow edges of ponds or lakes are often wetlands. Vegetation along the shore stabilizes soil, filters runoff to protect water quality, and provides habitat. Shade-providing trees moderate water temperature, which influences water chemistry and aquatic habitat.

Wetlands

Wetlands are land-water ecosystems where water saturation is the dominant factor—water affects soil and its characteristics, favoring the types of plants that are adapted to grow in wet conditions. The US Fish and Wildlife Service defines wetlands as ecological systems that are transitional between terrestrial and aquatic systems where the water table is usually at or near the surface or the land is covered by shallow water.[21]

In summary, three elements define wetlands:

- *Soil*: saturated or covered by water at some time during the growing season of each year
- *Water*: water table is at or near the surface
- *Indicator plants*: adapted to live in saturated soil

Though wetlands usually occur in low-lying areas, depressions, or other places that collect water or are connected to the water table, they can also be found on hilltops and slopes, along the edges of streams, rivers, floodplains, lakes and ponds, in fields and meadows, and in forested areas. Some wetlands contain standing water year-round; others are seasonally dry. Wetland types include marsh, fen, wet meadow, prairie pothole, vernal pools, and forested swamp. Each wetland type is a different ecosystem, with characteristic plants and animals. The US Fish and Wildlife Service classifies wetland types[22] in terms of their shared physical, chemical, and biological characteristics and has developed wetland maps based on habitat.[23]

The land that surrounds a wetland and drains into it is the wetland's contributing drainage area, which functions like a small watershed. Land use within this drainage area affects runoff and water quality, which in turn affects the wetland ecosystem. Wetland water supply comes from runoff, small tributary streams, groundwater, seeps, and springs, or from adjacent lakes, rivers, and ponds. Some wetlands recharge groundwater; others depend on groundwater as their water source. This movement of water between wetland and groundwater may be seasonal. A wetland may recharge groundwater in spring, when it receives more rain, and be replenished by groundwater discharge in fall, when there is less rain. Seasonal water-level changes in wetlands affect the system's nutrient cycling, microorganisms, plant and insect species, and fish and wildlife.

Wetlands play a critical role in regulating water movement through a watershed. Groups of wetlands work collectively as a system. They serve as "sponges" across the landscape, collecting water from precipitation or runoff until they become saturated, and then releasing it slowly. The amount of water a wetland can store depends on local conditions, wetland type, and soil permeability. The following are three examples of storage capacity:

1. A simulation model for Grant County, Minnesota, showed that restoring 25 percent of the farmed and drained wetlands within one drainage basin would increase watershed storage capacity by 27–32 percent, and a 50 percent restoration would increase storage by 53–63 percent. Wetlands have the potential to store up to 20 percent of the basin's total precipitation.[24]
2. In South Carolina, wetlands like pocosins and hardwood swamps store an estimated 45.8 billion gallons of water (enough to fill seventy thousand Olympic-size swimming pools).[25]
3. An Indiana study presents a general measure of wetland storage capacity—a one-acre wetland, one foot deep, can hold approximately 330,000 gallons of water.[26]

Wetlands can store water as soil moisture, groundwater, and surface water; storage capacity may fluctuate with the seasons. When wetlands are removed, watershed storage capacity is lost, and flooding increases.

Wetlands play other roles within a watershed that contribute to the watershed's ability to reduce flooding, protect water quality, and replenish water supply. Various wetland types, such as swamps, marshes, or wet meadows, produce different services depending on their vegetation, size, condition, region, and position in the landscape. Numerous small wetlands can contribute significant total wetland acreage within a watershed and increase its ability to protect

TABLE 1.3. Wetlands: watershed roles and examples of benefits

Wetland role in watershed	Examples of benefits
Water-quality protection and improvement	Filter and trap nutrients (e.g., nitrogen, phosphorus) and other pollutants from stormwater runoff; wetland plant roots and soil microorganisms absorb dissolved nutrients and facilitate chemical breakdown of pollutants (e.g., biological degradation and chemical oxidation)
Groundwater recharge	Collect surface water and let it seep slowly into the ground to recharge groundwater and aquifers (drinking-water supply)
Flood control and mitigation of climate-change effects	Absorb, store, and slowly release floodwaters resulting from heavy rainfall, reducing stream flows during periods of high water and maintaining base flows during drought
Sediment control	Provide natural settling basins for sediment in stormwater runoff; wetland vegetation binds soil particles and slows the transport of sediment
Food-chain support and nutrient cycling	Supply food and organic detritus that support fish and wildlife in adjacent waters; nutrient uptake, transformation, and export to other habitats
Habitat	Provide important nesting, breeding, feeding, migratory cover, and wintering habitat
Biodiversity	Support high diversity of species, including many rare plants and animals
Cultural values	Support recreation, tourism, education, natural areas, and open space

water quality, reduce flooding, support habitat, and replenish water supply. Wetland services and benefits are listed in table 1.3.

Buffers

Buffers are vegetated areas along the edges of streams, lakes, and wetlands that protect these water resources from adjacent land uses. The characteristics of a buffer—such as width, slope, type of vegetation, and specific location in the watershed—affect the level of protection provided for the adjacent water resource and the living things it supports. Buffers are critical to maintaining watershed health. Throughout a watershed, buffers provide wildlife habitat, stream-bank protection, water-quality improvement, and other benefits. Although wetlands and buffers may share similar roles within a watershed, they differ in how these roles are accomplished.

Table 1.4 presents a summary of buffer roles and benefits.

TABLE 1.4. Water resource buffers: watershed roles and benefits

Buffer role in watershed	*Examples of benefits*
Flood control	Intercept, slow, and absorb stormwater runoff; enhance wetland flood control, protect floodplains, facilitate gradual release of flood flows within watersheds
Nutrient and pollutant removal	Intercept and filter stormwater runoff; reduce or remove pollutants (including road salt, fertilizers, herbicides, pesticides, and heavy metals) from stormwater runoff before they reach wetlands, lakes, and streams
Habitat	Provide important wildlife travel corridors and wetland-to-upland transitional habitats vital to the survival of many wetland- and stream-dependent species
Bank stabilization and shoreline anchoring	Intercept erosive force of runoff, stabilize banks and stream channels, control or prevent soil erosion
Visual/noise barrier	Protect wetland-, lake-, and stream-dependent wildlife from human disturbances
Temperature regulation	Provide shade, particularly during the growing season, to maintain cool, well-oxygenated water during dry or warm periods
Detrital input	Provide organic (e.g., leaf litter) and woody debris, important sources of food and energy for fish and aquatic invertebrates
Cultural value	Preserve natural open space; provide areas for education, research, and passive recreation

Forested Areas

The extent of forested land within a watershed is another measure of watershed health. We measure forested areas in watersheds in terms of (1) the percent of the total watershed that is forested, or (2) the extent and width of forested cover along the edges of lakes, streams, and rivers. Forested cover in headwaters and along small tributary streams maintains healthy downstream river conditions.

Trees contribute to overall watershed health by

- *Reducing stormwater runoff and flooding.* Trees absorb rain and snowmelt, take up water from soil, slow runoff, and provide organic matter to soil. Forested areas produce little runoff; they recharge groundwater and aquifers and help to sustain stream flows.
- *Reducing erosion of stream channels.* Tree root systems stabilize soil, provide organic matter, and hold banks in place.

- *Improving water quality.* Trees filter sediment and some pollutants from stormwater runoff, take up nitrogen, and in some cases break down pollutants in soil and stormwater runoff.
- *Providing habitat.* Wooded areas in floodplains and along streams provide habitat for a variety of birds, mammals, reptiles, and amphibians. Trees along the water improve aquatic habitat by supplying organic matter for aquatic food chains and moderating (cooling) water temperatures.

A forest's tree species, age, and location influence these benefits in a particular watershed subbasin. In regions where watersheds have little or no forested cover, or where much of the watershed is developed, a strip of trees and shrubs along stream corridors is especially important for ecosystem and watershed health.

Groundwater and Aquifers

Streams, ponds, and wetlands are visible in the landscape. But most of our freshwater escapes our notice because it lies beneath the land surface. Rainwater and runoff move downward through empty spaces or cracks in the soil, sand, or rock. When this water reaches a layer of rock through which it cannot easily move, it fills the empty spaces and cracks above that impervious layer. The upper edge of this saturated soil is the water table. Sometimes, groundwater finds its way to the soil surface as seeps or springs, which may supply water to small streams, ponds, and wetlands. Pumping groundwater may impact the whole watershed because of the interconnections between groundwater and surface streams, wetlands, and lakes.[27] Figure 1.4 shows an example of this interconnection between well pumping and a stream.

Groundwater is replenished by (1) precipitation that percolates through the soil, and (2) water recharge from streams, lakes, and wetlands. The proportions of water that these sources contribute vary according to climate, geology, and region.

Under natural conditions, groundwater moves along flow paths from areas of recharge to areas of discharge at springs, streams, lakes, and wetlands. The contribution of groundwater to these systems varies widely, depending on topography, soils, and drainage patterns. According to hydrologists with the US Geological Survey, groundwater contributes an average of 40–50 percent of the water in small and medium-size streams.[28] This percentage may vary seasonally, depending on local conditions, reaching 100 percent during periods of dry weather. It is also higher in areas underlain by highly permeable

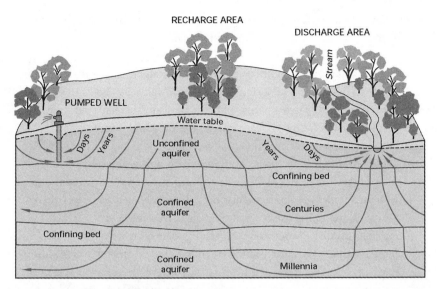

FIGURE 1.4. Groundwater flow paths. From Thomas Winter et al., *Ground Water and Surface Water: A Single Resource*, US Geological Survey Circular 1139, 1998. Courtesy of Department of the Interior, USGS.

sand and gravel deposits that allow for increased groundwater recharge. While percentages may vary, the important point is the direct connection between streams and groundwater.

An aquifer is a geologic formation composed of rock, sand, or gravel that stores a significant amount of water. Aquifers are often a source of clean drinking water for larger communities. The size of the spaces in the soil or rock and the connections between the spaces determine aquifer storage capacity. Sand and gravel form the most productive shallow aquifers, and carbonate rocks, such as limestone, form the best bedrock aquifers. The US Environmental Protection Agency defines a sole or principal source aquifer as one that supplies at least 50 percent of the drinking water consumed in the land area that overlies the aquifer. This land area may have no alternative drinking-water source.

Overall Watershed Benefits

The network of wetlands and streams, buffers, vegetated lands, and groundwater throughout the watershed provides collective watershed benefits. The value of a watershed lies in the health of this network and in the connections between land and water across the landscape, above and below the ground.

The following examples of watershed services and benefits include the collective benefits from wetlands, buffers, and forests:

- *Water supply.* Watersheds replenish groundwater, an important source for drinking water and agriculture. Groundwater also helps to protect aquatic habitats. During times of drought, watersheds maintain interconnections between surface water and groundwater that sustain surface waters like streams, lakes, and wetlands.
- *Water-quality improvement.* Wetland and buffer vegetation and soils filter pollutants from runoff. A network of forested areas, streams, and wetlands improves water quality and releases cleaner water to streams and lakes, reducing the costs of treating drinking water.
- *Flood protection.* Watersheds (especially their networks of wetlands) reduce vulnerability to flooding by collecting and storing water and releasing it slowly to lakes and streams. Floodplains absorb flood flows, slow surface runoff, and allow it to seep into the soil. Watershed systems naturally control stormwater runoff and erosion; vegetated surfaces in watersheds slow stormwater runoff, allowing it to sink into the ground or be taken up by vegetation.
- *Moderation of climate change.* Watersheds provide natural stormwater management that reduces vulnerability to increased flooding and other damaging effects of more severe and frequent storms. Watershed vegetation helps to moderate water temperature.
- *Biodiversity.* Watersheds support a variety of ecosystems and habitats (aquatic, wetland, and terrestrial) that together support plant and animal diversity, including rare species; they reduce vulnerability to invasive species and provide habitat connections via stream corridors, wetlands, and vegetated buffers.
- *Cultural values.* Streams, wetlands, and lakes provide opportunities for education and research, as well as water-based recreation and tourism. Watershed conditions can also affect property values: recreational water features, aesthetic amenities, water storage, water-quality improvement, and flood protection add to property value.

We can calculate the economic value of some of these benefits on a watershed-wide basis. For example, a survey of watershed characteristics and treatment costs conducted for twenty-seven drinking-water utilities found that for every 10 percent increase in watershed forest cover, chemical and other treatment costs decreased by 20 percent.[29] Developed areas with less forested (or other vegetative) cover must manage increased stormwater run-off, resulting in higher costs for stormwater facilities and infrastructure.

The Puyallup River community (in Washington State) compiled its watershed characteristics to "understand nature and the human economy as a single system."[30] "By reducing the frequency and severity of floods, supplying water, buffering climate instability, supporting fisheries and food production, maintaining critical habitat, enhancing recreation and providing waste treatment, among other benefits, the Puyallup River Watershed ecosystems provide between $526 million and $5 billion in benefits to the regional economy every year."[31]

WHEN a community identifies its natural ecosystems and watersheds and evaluates their services and benefits, it can use this information to dramatically improve the effectiveness of comprehensive plans, open-space plans, zoning, local ordinances, and conservation-easement management plans. Details about natural systems characteristics can be used to evaluate effects of land-use activities on water, a process we will explore in the next few chapters. The process begins with how we look at the land.

2

Picturing Environmental Features and Systems

Water is the most critical resource issue of our lifetime and our children's lifetime. The health of our waters is the principal measure of how we live on the land.

LUNA LEOPOLD, former USGS chief hydrologist

We see the land from a particular perspective. Whether it's the mountaintop view or the boots-in-the-water search for details, as we describe a particular site from various points of view we notice different features.

The thrill of looking down at the world from a high vantage point grips artists and adventurers alike. On a gorgeous blue-sky June morning, this "vantage point" attraction prodded me up New York's Shawangunk Ridge and led me sweating over the last rock faces to the top. Pushing laurel scrub out of my way, finding toe grips and handholds, I hoisted myself up to a great flat slab, let down my backpack, and took a long drink of cool, sweet water. I had a great view of the mid-Hudson valley: a rolling patchwork of forested and open land; shining stream-ribbons; and ponds, houses, and roads. I can picture the view now almost as a watercolor landscape—brushstrokes capturing the feel as well as its appearance. My home was down there, a six-acre parcel hidden among the trees.

The last thing on my mind was the science behind what I was seeing—watersheds, ecosystems, and nutrient cycles. I didn't think of the view in those terms at all. But the broad strokes that capture the landscape also cover myriad details of leaf and stream riffle, wing beat, and frog capturing fly. Identifying these details requires a different perspective, from ground—or water—level. Details fill in our picture and supply essential clues about how the systems we value are faring, and how human activities may affect them.

In this chapter, we will look at how to translate what we discover from these different perspectives into information that will help us identify impacts accurately, so that we can evaluate how land-use activities affect water resources. This approach begins with mapping and progresses to gathering more detailed information about soil, water, and living organisms. You may need a professional evaluation of an area's ecological characteristics to supply the field analysis and level of detail necessary to withstand expert scrutiny and legal challenges.

A thorough description of "existing conditions" helps the reviewer to get a baseline picture of the natural resources that various land development activities may affect. It's the first step in a longer process:

1. Describing ecosystems and watersheds and their services
2. Evaluating effects of land-use activities on these systems and services
3. Weighing tradeoffs, mitigation, and true costs

If "existing conditions" descriptions are wrong, missing, or incomplete, then you won't be able to match up impacts with specific changes to natural systems, or figure out which wetlands, streams, or lakes need protection. While this information may seem a bit technical and detailed, it is critical as the basis for a solid evaluation.

Descriptions of existing environmental conditions often only meet the regulatory minimum; in other words, we describe only as much as the law requires. If we leave out the details of how ecosystems and watersheds work, we may not evaluate some of the significant impacts on these systems and the benefits that they provide for us. The more accurately you can describe existing systems, the more solid your basis for describing how land-use activities change them. This chapter extends beyond a "minimum information" approach, builds on the concept of natural systems and what they need to remain viable, and challenges you to balance the details you gather with your view from the top of the mountain (or a good aerial photo). Sources for describing and understanding natural resources are numerous, available, and vary by region; some general national references can help you get started.

Mapping

With a good map in hand, you will know what to look for and can make sense of what you see on a particular site. Geographic Information Systems (GIS) tools and an array of existing maps make identifying the natural resources on a parcel considerably easier. Still, mapping is not perfect, so you will need to verify conditions in the field to make sure maps reflect what is actually on the site, and to capture specific features that don't appear on existing maps.

Watersheds and Subbasins

So where is your site within that watercolor view from the top of the mountain? Begin with a topographic map and the most recent aerial photography available. US Geological Survey (USGS) topographic maps are especially useful for this level of review. With no parcel boundary lines, these maps allow you to see a site in the context of the surrounding land and water features. Every parcel of land is located within a watershed. A building site may contain portions of several watershed subbasins. Check with your state and county governments or local agencies to discover detailed local watershed information. As a starting point, the USGS website, www.water.usgs.gov, includes a useful tool listed as "Science in Your Watershed," with maps to locate your watershed. For more detailed subbasin maps, refer to the USGS StreamStats interactive mapping system at that same website.

Large watersheds contain smaller watersheds, or subbasins. When you divide a larger watershed into smaller, more manageable areas, it is easier to identify and evaluate impacts on watershed characteristics. For example, in New York's Hudson River watershed, where I live, state and regional programs reflect concerns about the river's problems on a large scale. The Hudson River has sixty-five major tributaries; its watershed is 13,400 square miles. I cannot find my own property on a map at this scale. The watershed of the Wallkill River, a major tributary to the Hudson, is a 785-square-mile area within the Hudson River watershed. This is still too large an area to mean much to me and my six-acre parcel of land (Map 2.1).

One of the Wallkill River's sixty-nine tributaries is the Shawangunk Kill ("kill" is the Dutch word for stream, reflecting early Dutch settlement of the area), thirty-five miles long, which drains a watershed of 147 square miles (Map 2.2). The little stream that flows through the woods behind my house is the Dwaarkill, a tributary to the Shawangunk Kill.

The Dwaarkill is just over nine miles long, flowing from the Shawangunk Ridge down to the Shawangunk Kill. Its watershed is about eleven square miles; at that scale, you can locate a parcel or a house (Map 2.3). It's much easier to identify impacts within this eleven-square-mile area than within the larger watersheds in which it is nested. It's also easier to find the headwaters and identify the source of water for each small stream. When I think of water and its condition on that subbasin scale, I have a better feel for its personal value to me, and the local land-use activities that may affect it.

If you want to trace the flow of water and any pollutants it may be carrying, you begin at a point within a small subbasin and follow small tributaries to

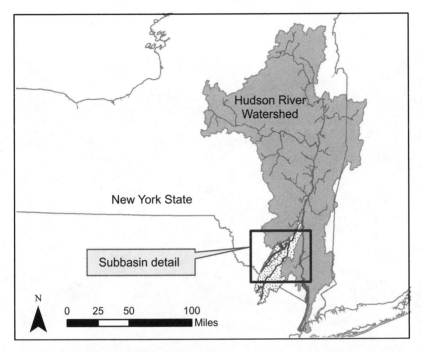

MAP 2.1. Hudson River watershed. Map by Angela Sisson.

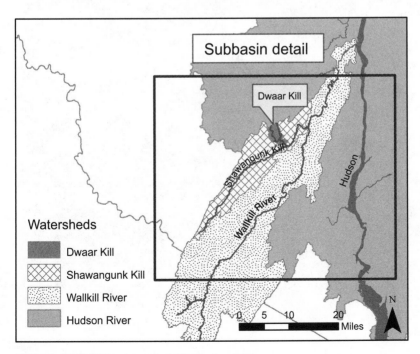

MAP 2.2. Wallkill River watershed. Map by Angela Sisson.

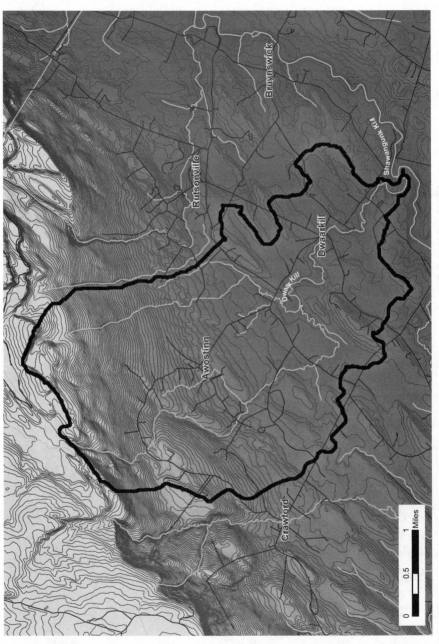

MAP 2.3. Dwaarkill subbasin. The Dwaarkill begins on the ridge and is fed by several tributaries before it flows into the Shawangunk Kill. All tributaries of the Dwaarkill are shown in white; roads are black. Map based on USGS StreamStats website information, adapted by

large tributaries that eventually lead to rivers. A river with all its tributaries gives you a good picture of how small streams can contribute contaminants and cumulatively cause water-quality problems downstream.

When you are evaluating environmental effects, rather than limit your view to a site-plan map, it makes more sense to look at all waters within a watershed or subbasin. Include all wetlands and streams (intermittent, perennial, and ephemeral), seeps, and springs—regardless of their regulatory status or jurisdiction. Consider belowground as well as surface water. If aquifer maps are available, identify any aquifers in the vicinity of the site you are evaluating. Access state and county websites for information about aquifer locations. A map of all water resources helps you get a clear picture of the movement and storage of water. Knowing this, you can track where water (or waterborne contamination) is coming from, where it's going, and what it passes through along the way. This picture can be compromised when some water features, such as small wetlands or streams, are left out.

We often identify wetlands in terms of regulatory standards—which wetlands are protected by regulations, and which agency is responsible. Because of federal and state gaps, many wetlands remain unprotected, leading to flawed review of impacts on water quality and supply. For example, a rapidly growing Hudson Valley town has a history of well contamination and water-supply shortages. The local planning team reviewed a site plan for a 150-acre residential subdivision in a wooded area. The site contained numerous small wetlands and streams, many of which didn't meet federal or state criteria for protection. Despite existing local water problems, the town attorney insisted on identifying and evaluating only "jurisdictional" wetlands for development-related impacts. This approach left out the interconnections between aboveground and belowground waters; and the flood reduction, groundwater recharge, and other services provided by "non-jurisdictional" wetlands.

The classification of small streams can generate similar confusion. In a different Hudson Valley town, where regulations protect perennial but not intermittent streams, site plan reviewers argued about definitions—that is, how many months out of the year a particular stream carried a "measurable" flow of water, and whether it was intermittent or perennial, according to regulatory criteria. No one mentioned the value of the stream for flood reduction or habitat, its connections with groundwater or downstream waters, or differences in flow due to drought or seasonal climate variations.

Many of us are interested in reducing local flooding, even along very small streams; watershed mapping should include floodplains. A good source for floodplain maps is the Federal Emergency Management Administration's website at www.msc.fema.gov, where you can look up floodplains and flood-hazard areas by address. Note local flood-prone areas on your watershed map.

Topographic Features

Watershed boundaries are formed by topographic features. Topographic and soil maps illustrate some of the connections between streams, lakes, and wetlands. Describe major geologic features within your watershed—including rock outcrops, gravel banks, caves, underground karst topography, caverns, and sinkholes. These features can be key to understanding local water systems. For example, karst is shaped by types of chemically soluble bedrock (limestone, dolomite, and marble); its features include sinkholes, vertical shafts, underground streams, and caves.

Topography influences drainage patterns, affecting how water moves across the land and where it collects. The contour lines on topographic maps can be used to identify small drainages (such as intermittent streams) that may not be mapped but that are important as part of the watershed's network of surface waters. Drainages usually show up as a series of nested contour-line V's; water flows from the high point of the V toward the open arms. Both slopes and low-lying areas indicate places that may be particularly vulnerable to runoff and flooding.

Erosion may be a problem on steep slopes with shallow soils. A slope of 15 percent or greater is a general guideline for "steep," though some soil types are vulnerable to erosion even on lesser slopes. For soil maps, including areas with steep slopes, use the US Natural Resource Conservation Service county soil survey maps or the Web Soil Survey at http://websoilsurvey.sc.egov.usda.gov/App/HomePage.htm.

Wetlands

Mapping wetland boundaries can be challenging, depending on the level of detail and accuracy required. Approximate wetland locations based on topography, aerial-photo interpretation, and existing soil and wetland maps may be sufficient for an overview of wetlands within a watershed. You can get a good idea of approximate wetland locations by using the US Fish and Wildlife Service's National Wetlands Inventory (NWI) maps, available online with a user's guide at http://library.fws.gov/wetlands/nwimaps_madeeasy.pdf. Based on wetland habitat, these maps identify many, but not all, wetlands. Hydric soils found on soil maps indicate areas where wetlands are likely. To locate wetlands, use a combination of existing maps, including the NWI wetland habitat maps, topographic maps, aerial imagery, soil maps, and locally available wetland maps.

NWI-mapped wetlands are not the same as those regulated by law under the US Army Corps of Engineers. No corresponding map shows wetlands regulated by the Corps; each must be identified and delineated individually, in the field.

Wetland mapping provides details necessary for accurately evaluating impacts, identifying important habitat, describing overall watershed condition, or determining appropriate buffer location and size. For these reasons, you should identify and map all wetlands regardless of regulatory jurisdiction on the parcel or site you are evaluating. Your map should include the entire wetland, whether or not it extends beyond parcel or municipal boundaries. Along with each wetland, sketch in its contributing drainage area (CDA). The Center for Watershed Protection offers detailed instructions for mapping a CDA.[1]

Local, state, or federal law may require wetland delineation and the onsite mapping of actual wetland boundaries. Regulatory guidelines set criteria to identify and map wetlands. Specific characteristics of wetland soils, water, and plants are the basis for identifying wetlands in the field and for determining whether or not a law protects a wetland. Definitions guide consistent delineation of wetlands based on Section 404 of the Clean Water Act. Only wetlands that meet "404" criteria fall under federal regulatory jurisdiction. The US Army Corps of Engineers 1987 Wetland Delineation Manual, and the Regional Supplements released in 2012, describe wetland indicators and accepted delineation and mapping practices. The 2012 National Wetland Plant List is a complete illustrated listing of wetland plants by region, found online at http://wetland_plants.usace.army.mil.[2]

Many states rely solely on federal wetland protection regulations and the US ACE 1987 Wetland Delineation Manual for delineating wetlands. States and municipalities that have adopted their own wetland regulations may have additional mapping information and GIS layers available for your use. The Environmental Law Institute website (http//www.eli.org) includes an overview of state wetland programs. Additional information is available from the Association of State Wetland Managers (http://www.aswm.org) and from the US Environmental Protection Agency (http://www.epa.gov/owow/wetlands/initiative/).

"Delineating" a wetland requires identifying its boundaries, placing flagging along those boundaries, and transferring boundary data to a site map. Wetland boundaries aren't always obvious; many wetlands are dry seasonally or during drought. Delineations should be conducted during the growing season for optimum accuracy.

Wetland delineations can be controversial when they identify areas subject to regulatory protection and affect the location of development activities. Sometimes wetlands are missed altogether or under-delineated (that is, the mapped boundary is smaller than the actual wetland boundary). Property owners who may not understand delineation rules and procedures may assume that wetland boundaries can easily be changed to accommodate site plans.

For example, in Breckenridge, Colorado, the site for a large residential development included an extensive wetland area with a mosaic of different vegetation types. While delineating these wetlands, I climbed a hill onsite for an overview of the vegetation patterns. Soon the developer showed up. He anxiously unfurled a large site-plan map and pointed to an upland area outside the condos and their access roads. He said: "Can you put the wetland here? I need this area"—he pointed to the wetland—"for the parking lot." That request got my attention. "Well, that's not really how it works," I began, wondering how I could put this diplomatically. "I have to follow a specific method—wetlands are where they are," I told him. "I can't change that."

Delineators are often the only authority for determining where the wetlands are, and whether they fall under state or federal regulatory jurisdiction. Staff cutbacks in state and federal regulatory agencies reduce oversight for wetland-boundary verifications. In ambiguous field cases, because of variations in wetland training and differences about drawing the boundary line, it is often advisable to have a third party verify a delineated wetland to make sure the boundary is accurate.

Habitat Mapping

The map you've been compiling should show you where the wetlands, streams, and lakes are located; now you can add ecosystem information by mapping habitats. The first step is to figure out where all the different types of terrestrial, aquatic, and wetland habitats are located—on and adjacent to your site. Although we make most project decisions on a site-by-site basis, wildlife, like water, doesn't stay put within parcel or municipal boundaries. Within the course of their life cycles many species use several different habitats. Land disturbance that avoids breeding habitat, for example, may unintentionally destroy adjacent feeding sites or winter habitat. To include all ecosystem connections, it's useful to place habitat within a watershed or subbasin context. Sketch habitat areas onto the map you've already generated so you can evaluate how they relate to topographic features, watersheds, soils, roads, and developed areas. On the map, identify all wetlands, lakes, and small streams (intermittent and perennial) as habitat types.

A habitat map shows the relative position of habitats, the size of each habitat area, and the degree to which habitat fragmentation may already be a problem. Habitat fragmentation breaks large contiguous areas into smaller areas that are no longer connected. It can interfere with ecosystem processes, disrupt the movements of wildlife species between habitats, and result in habitat loss.[3]

Several tools can help with habitat mapping. The Cowardin system of classifying wetland and aquatic habitats describes units for mapping and establishes a consistent method for identifying habitats by plants, soil or substrate, and hydrology. The National Wetland Inventory uses this system, and you may want to refer to it as a guide to your mapping of stream, wetland, and lake habitat types.

You should estimate how much of the watershed or subbasin is forested and the extent of vegetated buffers, including forested edges along rivers and streams. GIS and aerial imagery are useful tools. The National Land Cover Data found on the EPA's website is another useful resource for estimating vegetation cover.[4]

Habitat and species data are available from your state's Natural Heritage Program and from other regional or local sources. For example, in New York State, *Ecological Communities of New York State* is a useful resource;[5] the *Biodiversity Assessment Manual for the Hudson River Estuary Corridor* presents detailed habitat and species information.[6] The latter is a user-friendly model for evaluating habitats that could be applied to other geographic regions.

Use the system (or combination of systems) that works best in your situation. Exploring your site on foot and observing habitat firsthand is still the most accurate method, but it requires some expertise. You will need to field check all habitat data to verify type and location and to map small areas like turtle nesting sites that don't correspond to vegetation communities. You may want to recruit a biologist or ecologist to make more detailed habitat maps.

Land-Use Overview

Aerial photos and GIS mapping can picture the characteristics associated with developed land. Constructed features that change the movement of water include ditches, irrigation infrastructure, reservoirs, dams, buildings, roads, and stormwater-management infrastructure.

TABLE 2.1. Estimated impervious cover by land-use type. From *New York State Stormwater Management Design Manual*, and K. Cappiella and K. Brown, *Impervious Cover and Land Use in the Chesapeake Bay Watershed* (Ellicott City, MD: Center for Watershed Protection, 2001).

Land-use category	Mean impervious cover (%)
Agriculture	2
Open urban land (developed parks, recreation areas, golf courses, cemeteries)	9
2-acre lot residential	11
1-acre lot residential	14
½-acre lot residential	21
¼-acre lot residential	28
⅛-acre lot residential	33
Townhome residential	41
Multifamily residential	44
Institutional (churches, schools, hospitals, government offices, police and fire stations)	31–38
Light industrial	50–56
Commercial	70–74

Land development increases the extent of impervious surfaces that water can't penetrate—such as residential rooftops, patios and driveways, roads, buildings, and parking lots. Remote sensing and aerial photography are two accurate methods for measuring or estimating the existing impervious cover;[7] several other methods can be found online.[8] Impervious cover is also directly related to land-use or zoning categories, as seen in table 2.1.[9]

An estimate of existing impervious surfaces is a baseline to compare with new impervious surfaces. For these comparisons, make sure to use a consistent method for estimating impervious cover, as results may differ among the various methods. Impervious surfaces influence water quality and supply and play a significant role in the evaluation of land-use impacts, mitigation, and cumulative effects, discussed in later chapters.

Filling in the Details

Remember the view from the top of the mountain (an aerial photo can serve the same purpose) as you consult your maps and gather information. Maintain a balance between the two perspectives—use overview as well as details for a balanced approach to understanding the water resources on your site. Conduct a site visit after you have identified site features and water resources using maps, GIS, and other available tools. Use your observations to verify or modify your map(s) and to fill in some details about natural systems. Maps are not a perfect replica of what you may find in the field. They require interpretation so you can answer questions such as "Which wetlands provide maximum flood protection?"; "How is a stream connected to groundwater?"; and "How should land activities be managed or limited on a particular site to protect water quality?"

Depending on the size of the site and the complexity of the issues regarding its use, you may need a team to interpret maps and conduct site visits.

Details can overwhelm as well as inform; focus on the ones that give you critical information about managing or preserving water resources. To be effective in your efforts, pay attention to the details that matter most, and forge ahead even though you may not have every last bit of data. No one "knows it all" about the environment and its physical, chemical, and biological properties. We all have more to learn. To help you sort through the details, the following discussion zooms in on several points of particular concern about soil, water resource buffers, wetland hydroperiod, stream biomonitoring, and habitats.

What's Underfoot?

Soil is the foundation for upland and wetland ecosystems at the edges of aquatic habitats. The size of mineral particles determines the soil texture as a proportion of clay, loam, or sand. Soil is made up of particles of different sizes that may be mineral (from the breakdown of rocks) or organic (from dead plants and animals). When soil is dry, air fills the spaces between the particles; after rain, these spaces fill with water. Depending on soil type and climate, soils contain varying proportions of water and air. Soil characteristics influence vegetation. If you have ever planted a garden, you know how soil can affect the plants you grow. To produce the best fruit, vegetable, or flower you may have to add compost, nutrients, sand, or other soil amendments. Different plants in your garden have particular soil requirements, and the same is true in nature.

Soils are living systems. Though most of their workings are hidden from our view, what goes on underground affects what we see aboveground. Depending on type and location, soils contain a variety of organisms that maintain this "foundation" ecosystem, including live plant roots, microorganisms, beetles, worms, and larvae that help aerate the soil and cycle nutrients. An amazing world can be found in "one cubic foot" of soil, as described in the February 2010 *National Geographic*:

> Together with the bacteria and . . . microorganisms swimming and settled around the mineral grains of the soil, the ground dwellers are the heart of life on Earth. . . . The terrain they inhabit is not just a matrix of dirt and rubble. The entire ground habitat is alive. Living forms create virtually all of the substances that flow around the inert grains. If all the organisms were to disappear from . . . [the soil] . . . the molecules of the soil or streambed would become smaller and simpler. The ratios of oxygen, carbon dioxide, and other gases in the air would change.[10]

Incredible organisms live in soil. They are critical for the soil's ecosystem, influencing water and nutrient cycles, plants and their root systems, and aboveground ecosystems.

Information about soil types and their characteristics, including maps and tables, is available through your local agricultural extension service office or online at the US Department of Agriculture's Natural Resources Conservation Service (NRCS) site (www.ncrs.usda.gov/), or the Web Soil Survey also at the NRCS site. Soil surveys, available for most of the country, supply the details about soil characteristics. Table 2.2 presents a sample from the Soil Survey of Ulster County, New York, of soil characteristics that are related to water. Because soil survey maps are compiled on a large scale, soils on a particular site may need to be field-verified to confirm soil types and characteristics.

TABLE 2.2. Selected soil characteristics. Soil information from the Soil Survey of Ulster County, NY. US Department of Agriculture.

Soil series and map abbreviations	% slope	Depth to bedrock (inches)	Depth to water table (feet)	Drainage class	Hydro-logic group	Limitations for septic tank absorption fields
Bath gravelly silt loam (BgC)	8–15%	>60	2.0–4.0 perched	Well-drained	C	Severe; percs slowly
Farmington (FAE)	25–35%	10–20	>6.0	Well-drained	C	Severe; slope, depth to rock
Hoosic gravelly loam (HgB)	3–8%	>60	>6.0	Somewhat excessively drained	A	Slight*
Riverhead (RvA)	0–3%	>60	>6.0	Well-drained	B	Slight*
Wayland (Wb)	<1%	>60	0–0.5	Hydric; poorly drained	D	Severe: wetness, floods, percs slowly

Possible pollution hazard to streams, lakes, or springs, or to underground water supplies because of rapid permeability or fractured bedrock.

Table 2.2 lists Wayland soil as hydric and poorly drained, indicating the likely presence of wetlands. In this soil type, the water table is at or within six inches of the soil surface. The depth to water table is the depth where water completely saturates the soil. In wetlands, the water table is often very close to, or even at, the surface. The water table may fluctuate seasonally. Bath soils have a "perched" water table, which rests on top of an unsaturated layer of soil or rock. In some places, a dry zone separates an upper or perched water table from a lower one.

The water table may lie above a pan, which is a compact, dense layer in a soil that impedes the movement of water and restricts the growth of roots. Types of "pans" include hardpan, a hardened or cemented soil; fragipan, a loamy and brittle soil that appears to be cemented and restricts root growth; and claypan, a dense, slowly permeable layer with a high clay content that is usually hard when dry and sticky when wet.

The Natural Resource Conservation Service classifies soil types into four hydrologic soil groups based on the soil's runoff potential and infiltration—the rate at which water filters through the soil. Hydrologic group characteristics control aquifer-recharge rates and suitability of the soil for sustainable septic systems.[11] Group A generally has the lowest runoff potential and highest infiltration rate, and D the greatest runoff potential and low infiltration rates. If part of the acreage is artificially drained and part is undrained, a soil may belong to two groups.

The last column in table 2.2—Limitations—presents the suitability of soil types for various land uses. For example, some soils are poorly suited for septic absorption fields. Of the soils listed on the table, one is a poorly drained wetland soil, another has a perched water table, and one has a shallow depth to bedrock and steep slopes. Soils with "severe" limitations have unfavorable properties requiring special design, increased maintenance, and sometimes significantly higher construction costs. Well-drained Hoosic and Riverhead soils are more suitable for septic absorption fields, but because they have high infiltration rates, they also have the potential to carry pollutants from the surface to groundwater. "Soil limitations" illustrate how soil characteristics can influence road building, septic systems, and construction of basements.

To learn more about soils, including chemical characteristics, habitat, and agricultural suitability, refer to the soil surveys or the Web Soil Survey.

How's the Water?

A map of a subbasin and its surface waters gives you a starting point for understanding how the water systems on your site are connected. Details about water features help you understand how these systems work and how you can protect them. The following discussion about buffers, wetland hydroperiod, and stream biomonitoring will help you fill in some of those details.

Buffers

Specific stream, lake, and wetland conditions determine the overall health of a watershed or subbasin. A healthy wetland, stream, or lake includes a vegetated buffer. You can estimate the extent and condition of existing buffers from aerial photos and verify them during field visits.

The ideal size of a buffer depends on its purpose. For example, for water quality protection, buffers need to be adequately vegetated and wide enough to process the contaminants in runoff before they reach the water. If vegetation is inadequate for holding soil in place, rivulets and eroded channels may develop, impairing the water-quality protection function of the buffer. Several conditions influence buffer effectiveness:

- *Purpose.* Specific buffer characteristics are necessary for the achievement of a particular buffer function or service. Is the main function of the buffer wildlife habitat, water-quality protection, or stream-bank stabilization?
- *Vegetation type and condition.* Buffer vegetation (preferably native species) can slow or intercept stormwater runoff. Slower water drops some of its sediment load, causes less erosion, filters contaminants, and allows runoff to seep into the soil.
- *Extent.* Ideally, a buffer surrounds a lake or wetland or extends along both sides of streams. Buffered edges of stream headwaters and small streams protect habitat, water quality, and water supply in downstream waters. Buffers that are patchy rather than continuous—especially where adjacent land is densely developed—are less effective.
- *Intensity of adjacent land use.* Effective buffers lie directly between the water's edge and sources of contamination or disturbance on adjacent land. Runoff from a large impervious surface or an area with a high chemical load (fertilizers and pesticides, for instance) requires a larger buffer for water-quality protection.

No single size for buffers will fit all circumstances. To compensate for more rapid runoff, buffers on steep slopes or rocky soils need to be wider. Particular wildlife species need specific minimum buffer widths for habitat protection. Buffers must be a minimum width to successfully filter certain types of pollutants from runoff.

Guidance for determining buffer size comes from a variety of studies and resources. In 2002, the Ecological Society of America gathered leading scientists to develop guidelines for land-use decision making. Based on these guidelines, the Environmental Law Institute (ELI) surveyed scientific literature to describe ecological thresholds for protection of natural resources.[12] The survey on riparian buffers reviewed more than 140 recommendations. The ELI study recommended these buffer widths:

- 25 meters (82 feet) for nutrient and pollutant removal
- 30 meters (98 feet) for temperature and microclimate regulation
- 50 meters (164 feet) for detrital input and bank stabilization
- 100 meters (328 feet) and more for protection of wildlife habitat, depending on the species and specific habitat needs (narrower areas can still function as wildlife corridors)

The Environmental Law Institute recommends that "land use planners should strive to establish 100 meter wide riparian buffers to enhance water quality and wildlife protection" and that watershed protection efforts should establish continuous buffer strips along streams and protect stream headwaters as well as broad downstream floodplains.[13]

A 2001 Massachusetts study by Lynn Boyd documents wetland buffer use by reptiles, amphibians, birds, and mammals.[14] The study emphasizes that while it is difficult to establish a "blanket width" appropriate for all purposes, the importance of providing full protection to at least one hundred feet beyond the wetland is evident. Some 77 percent of Massachusetts's wetland-dependent wildlife relies on buffers that are at least one hundred feet wide. More than half the state's wetland-dependent species also depend on habitat areas beyond one hundred feet from the edge of a wetland.

Forest cover both within buffers and throughout a watershed is an important component of watershed health. It can be measured as a percent of total watershed area, or as linear riparian buffer. There is no one ideal value for percent forested cover in a watershed, though research provides examples:

- At least 65 percent watershed forest cover is needed for the presence of a healthy aquatic insect community in Washington's Puget Sound.[15]
- In Montgomery County, Maryland, remote sensing studies measured tree cover, impervious surfaces, and riparian buffers and rated stream health as "excellent" when at least 65 percent of the length of the stream network in the watershed is forested (within one hundred feet of the

stream's edge). Streams with at least 45 percent cover achieved a rating of "good."[16]

The Trust for Public Land and the American Water Works Association added an economic component, finding that more forest cover in a watershed can lower the treatment costs for drinking water. According to the study, forest cover up to about 60 percent resulted in lower water-treatment costs; for every 10 percent increase in forest cover, treatment costs decreased by about 20 percent, leveling out with forest cover at 70–100 percent.[17]

Additional references, which may be printed or available as PDFs online, contain specific information about buffers and forested cover:

- *Conservation Buffers: Design Guidelines for Buffers, Corridors, and Greenways*[18]
- *Planner's Guide to Wetland Buffers for Local Governments*[19]
- *Riparian Buffer Width, Vegetative Cover, and Nitrogen Removal Effectiveness*[20]

Wetland Hydroperiod

Information about a wetland's water source and its seasonal fluctuations (hydroperiod) helps us identify the degree to which land-use activities affect wetlands. To determine whether the wetland is working as a groundwater recharge or discharge area, identify its source of water. Some wetlands are sources of water for lakes or streams; others are "sinks" where water flows in but doesn't flow out. And some wetlands have water flowing both in and out, via surface or groundwater.

Changes in groundwater availability and water-table levels may disrupt wetland hydroperiod. Each wetland type has a corresponding hydroperiod—sometimes obvious (for instance, a pond with four feet of standing water year-round) and sometimes subtle (wet meadows that are dry at the surface for most of the year). Changes in hydroperiod that are not part of a normal seasonal cycle may impair wetland functions (such as habitat and flood storage) and alter the mix of wetland species. Tiner's *Wetland Indicators*[21] is a good source of sample hydroperiod graphs that correspond to different wetland types. Table 2.3 shows examples of hydroperiod categories and their characteristics.

Wetland plants have different water-level requirements. Hydroperiod determines the type of plants that can thrive in a particular wetland. These plants in turn shape the habitat for animal species. Many wetland plants have narrow

TABLE 2.3. Overview of variations in wetland hydrology. Based on information from Ralph Tiner, *Wetland Indicators: A Guide to Wetland Identification, Delineation, Classification and Mapping* (Boca Raton, FL: CRC Press, 1999), 32.

Wetland hydroperiod	Characteristics
Permanently flooded	Wetlands covered by shallow water
Semipermanently flooded	Inundated throughout the growing season in most years
Seasonally flooded	Surface water present for extended periods throughout the growing season; dry by late summer
Temporarily flooded	Surface water present for brief periods (often two weeks or less) during growing season
Saturated	High water table and saturated soil during growing season; standing water rarely present
Seasonally saturated	High water table at or near surface in winter and early spring

ranges of tolerance for different water depths; a change of even a few inches can make a difference in where they grow. The National Wetland Plant List[22] "Wetland Indicator Status" describes plants according to specific water needs and helps define wetland presence. This national list of wetland plants by species and their wetland ratings is used extensively for wetland delineations and the planning and monitoring of wetland mitigation and restoration. Wetland plant indicator status, as noted on this list, falls into several categories:

- *Obligate*: Plants that almost always grow in wetlands.
- *Facultative wetland*: Plants that usually occur in wetlands but may occur in non-wetlands. These plants are usually found with hydric soils, often in sites where water saturates the soil or floods the soil surface at least seasonally.
- *Facultative*: These plants can grow in a variety of different habitats (wetland and non-wetland) depending on site conditions.
- *Facultative upland*: Plants usually occur in non-wetlands, but may occur in wetlands.

Depending on water-table depth, the type of plants growing in a wetland may vary seasonally. "In some wetlands, species with wetter indicator statuses may predominate earlier in the growing season, with drier site species dominating in late summer and fall due to seasonal differences in near surface wetness."[23] The presence or absence of certain wetland plants can also signal environmental conditions, including the wetland's source of water (groundwater or surface water), soil type, and water chemistry (for example, nutrients, salt, pH).[24] Changes in hydroperiod may favor nonnative plants that can tolerate higher or lower water depth, or widely fluctuating water levels.

Stream Biomonitoring

While streams on your map may all appear as identical blue lines, each stream looks different "in person." Streams may be wide or narrow, shallow or deep, with steep banks or broad floodplains. Their banks may be heavily vegetated or barren and eroded. To evaluate the overall health of a stream or river, you can use "rapid bioassessment," a method developed by the US Environmental Protection Agency for describing physical, chemical, and biological characteristics of streams.[25] By collecting and evaluating baseline data, you can monitor stream health over time and evaluate impacts as land-use activities change a stream's characteristics.

Insect larvae and other invertebrates ("macroinvertebrates") found on the stream bottom indicate water-quality conditions. Some species occur in healthy streams; others signal degraded water quality. You can derive reliable measures of water-quality and stream conditions by analyzing the relative abundance of different species of aquatic insects and other small aquatic organisms. A "biological assessment profile" based on that data describes water quality (from polluted to pristine) on a scale from zero to ten. While local groups can be trained to collect stream samples, someone who is experienced in using this method should interpret biomonitoring data.

The following sample of stream information collected under this protocol is arranged according to the field-data checklists in the *Rapid Bioassessment Protocols* appendixes.

Biomonitoring Field Data—Selected Characteristics:

Physical

- Surrounding land use
- Bank condition
- Width of stream
- Depth
- Current or flow (cubic feet per second, cfs)
- Substrate / sediment (percent rock, sand, silt, gravel on stream bottom)

Chemical

- Temperature
- Dissolved oxygen
- pH
- Turbidity
- Specific conductance

Biological

- Canopy cover
- Aquatic vegetation
- Algae
- Macroinvertebrates (snails and clams, and aquatic insects)
- Fish

Twenty-five states have adopted some form of stream assessment methodology to evaluate state aquatic resources.[26] Check with your state for its method. For streams with operating USGS stream gauges, you can access stream-flow data as well. Online (water.usgs.gov and waterdata.usgs.gov) you can find and map information about streams in your watershed. USGS supports the National Water Information System, the National Streamflow Information Program, WaterWatch (flooding data), StreamStats (subbasin and stream data), and Statewide Streamflow tables.

What Lives Here?

The presence of plants and animals in streams, forests, lakes, and wetlands breathes life into maps and printed materials. The appeal of pulling on a pair of boots and "seeing what's out there" can lead to discoveries that you may not find in reports or websites. Individually or as groups, living organisms sustain ecosystem health; they affect, and are affected by, biological, physical, and chemical characteristics of their habitat.

Habitat and species information can help you

- evaluate the overall health of living systems within your watershed;
- identify important stream and wetland corridors that connect habitats;
- present details about ecosystem benefits;
- assess biodiversity and the effects of land-use activities; and
- develop ecosystem mitigation and restoration.

At any particular site, identifying animals and their habitats can be a daunting task. Some organisms live in the mud, some in the water or in the air above the water; some live in the fringe of vegetation along the water's edge. Others move among different habitats seasonally or during various life cycles. Though plants don't move, their appearance changes every season as they emerge, flower, go to seed, or lie dormant. So how do we describe the organisms that live on a particular site?

From my view of the mid-Hudson valley at the beginning of this chapter, the six-acre property where I grew up was a dot in the distance. Zooming in, that small piece of land lies within the Dwaarkill subbasin (see map 2.3), at the edge of the stream's floodplain. The parcel contains a small pond and wetland complex. I remember how it looks; growing up, I had mucked in and around that pond and its squashy edges, catching frogs, turtles, and newts wriggling in the "waterweeds." I was curious about how many different kinds of animals I could find. At eleven I began to make "The List," and worked my way through pickerel frogs, spotted salamanders, painted turtles, and even water snakes. Birds stretched my frame of reference; I knew the great blue heron and wood duck—they were easy to identify. The great parade of bugs was impossible to catalog. Once I really started to look, insects were everywhere, and they were all different: from mud crawlers to the dazzling jewel-flash of dragonflies. The immense task became overwhelming, and I abandoned The List for the simpler pleasures of catching frogs and turtles.

When I was eleven, I had never heard of an ecosystem. Although the variety of life in the pond left an indelible impression, I didn't know how to organize what I was seeing and make sense of the pond as a system until years later. Today I know that the small wetland complex includes a spring-fed shallow pond with emergent wetland along its edges; an adjacent shrub wetland; and forested wetland with a vernal pool. Each has a different hydroperiod and dominant plant species. The variety of plants and habitats is home for an incredible variety of frogs, salamanders, turtles, insects, and birds; some of them, like the spotted turtle, are species of conservation concern. That pond shaped my career, and I never forgot what I learned about the water and its life.

How can describing a valuable ecosystem influence local land-use decisions? In Milan, a small town in New York's Hudson Valley, a planning team led by the chair of the planning board developed an approach for evaluating habitat and species as a routine part of site plan review. Frustrated with lack of habitat protection when development projects skipped analysis of habitat impacts, the team developed local policy guidelines. Milan's Habitat Assessment Guidelines streamlined the environmental-review process, made it more consistent from project to project, and presented an organized approach for evaluating impacts on habitats, species, and ecosystems. Several nearby towns have adapted this approach to fit their review needs.

The goal of habitat assessment is to show the relationship of plants and animals to one another and within a larger system. When applied to water resources, this approach recognizes the value of streams, lakes, and wetlands as living systems, not just physical water conveyances. For example, the stream biomonitoring protocol mentioned earlier in this chapter illustrates how living organisms indicate the overall health of the stream.

Habitat-Assessment Approach

After mapping the site's habitats, the next step is to organize basic habitat information. You may consult with someone who can identify plants that are characteristic or dominant for each habitat type. Table 2.4 is a sample presentation of wetland and aquatic habitats. It includes the area of the habitat found on the site as well as the total habitat area, onsite plus offsite.

Species Associated with Habitat Types

If suitable habitat is present and in good condition, the habitat assessment process assumes that the plants and animals known to use that habitat type may be present, or may use it at some stage of their life cycles.

You can confirm species presence by observing plants and animals at or near the site and in adjacent or connected habitats. However, many species, especially those of conservation concern, may be hard to find, and you may need to hire a professional biologist to conduct field surveys. Special surveys require specific methods conducted at a particular time of day or season, depending on the species; you may need multiple surveys over several years to find species that tend to move between habitats over time.

TABLE 2.4. Sample site summary of wetland and aquatic habitat

Habitat	Total habitat area	Onsite area	Characteristic species	Hydroperiod or streamflow	Connections with other water features
Forested wetland	10 acres	3.5 acres	Red maple (*Acer rubrum*)	Temporarily flooded	No inflow or outflow; located within stream floodplain
Emergent wetland	3 acres	2 acres	Common cattail (*Typha latifolia*)	Seasonally flooded	Spring-fed; outflow is small intermittent stream that flows into a small pond
Stream: aquatic habitat with cobble/gravel bottom	Total length: 9 miles	Onsite average width: 12 ft., length: 800 ft.	Mayfly (*Baetis spp.*), brook trout (*Salvelinus fontinalis*)	Perennial; water flows most of the time	Wetlands at headwaters; stream has two intermittent tributaries and flows into river

If you rely entirely on general field visits to observe an animal on the site (different from a species-specific survey), you may be tempted to conclude, "If we didn't see it, then it isn't present." The problems with this assumption are (1) you are less likely to see rare species unless you are specifically trying to find them (which means you may be limited to a certain season or time of day to get any results), and (2) it can be hard (and expensive) to prove that a species associated with a particular habitat is *not* present and does *not* use that habitat at some stage of its life cycle—without conducting lengthy and expensive field surveys during specific seasons and over a period of several years.

Your best bet is to list both: observed species and species that may be present based on suitable habitat.

From your "master list" of species that use or may use the habitats on your site, identify the following:

- *Species of conservation concern.* These include threatened or endangered species (state and federal), rare species, Species of Greatest Conservation Need (listed in state wildlife management action plans[27]), and Partners in Flight listed birds. Consult state Natural Heritage Program lists, breeding-bird atlases, and other regional, state, and local listings available for your location. Identify habitat needs for these species. Once you know which species of concern may use the habitats under consideration, you can request targeted field surveys to search for them. For example, salamanders that breed in vernal pools frequent these pools as adults only briefly during breeding season in early spring. After the eggs hatch, the aquatic salamander larvae live in the pools for two to three months as they mature and eventually develop lungs. The young salamanders emerge from the pools and spread out into the adjacent woodlands where they spend most of their adult lives.[28]
- *Species with key ecosystem roles.* These include common species important for ecosystem functioning (for example, nutrient cycling); keystone or other species (such as predators) critical to the health of the system; indicator species that by their presence or absence demonstrate ecosystem condition or characteristics. Note plants and animals that require specific habitat conditions for their survival.
- *Invasive species.* These include nonnative invasives that present a problem, for example plants that crowd out and replace native vegetation. You can find lists of these species online at the US Department of Agriculture's National Invasive Species Information Center (http://www.invasivespeciesinfo.gov). Also, check your state's natural resources sites and invasive species listings.

Species Information

Table 2.5 is a sample summary of species and their habitat needs. On your site map, you can associate habitats with species that may use them.

Many animal species that use a stream or wetland don't stay there all the time. Dragonflies and mayflies have different life-cycle stages that carry them outside the water during certain seasons. "Special requirements" may include food, special habitat conditions, and habitat variety. The spotted turtle, which may spend the winter in a red maple swamp, moves into a vernal pool in June, lays eggs in a nearby sandbank, and crawls into other wetland and woodland habitats during late summer and early autumn. All these habitats, and the connections between them, are necessary for the spotted turtle to thrive.

Habitat Condition and Quality

Habitats differ in their condition and quality. Some have been degraded or contaminated; others are too small to support some species. Habitat condition dictates which species you are likely to find; for example, species that require clean water are unlikely to live in a contaminated stream. In general, some characteristics to consider when evaluating habitat condition include size and connections with other habitats, degree of human disturbance, presence of invasive species, condition and size of vegetated buffers, and water quality.

Habitat Profile

To complete your site evaluation, put together a brief description of each habitat type and associated species of conservation concern. Refer to chapter 1 or previous sections on water resources to fill in specific information for wetlands, and use the "rapid bioassessment protocols" for stream habitats. If you can't find the necessary information for a particular perennial stream, request

TABLE 2.5. Sample summary of species and their habitat needs

Species of conservation concern	Habitat type(s)	Special requirements and food sources
Brook trout (*Salvelinus fontinalis*)	Perennial stream	Cool, clean water with high dissolved oxygen content; aquatic insects
Red-shouldered hawk (*Buteo lineatus*)	Mature woodlands near rivers or streams, hardwood swamp, wooded areas in floodplains, riparian areas and wooded edges of wetlands	Nest sites near water; food is small mammals, birds, reptiles, amphibians, and crayfish
Spotted turtle (*Clemmys guttata*)	Wet meadow, sandbank for nesting, hardwood swamp, pond	Variety of connected good-quality habitats for nesting, feeding, hibernating

it. Conducting a basic bioassessment doesn't have to be prohibitively expensive, but it does require specific equipment and training.

Knowledge about habitats, species, and ecosystems is essential for evaluating environmental impacts, restoring damaged or degraded sites, and setting aside protected areas. In this context, it may be useful to assemble data about specific habitats within larger systems. For example, the value of a stream corridor includes aquatic habitat and organisms, riparian and floodplain plants and animals, adjacent wetlands, and nesting habitat for migratory birds.

THIS chapter has provided an overview of methods for mapping and describing ecosystems and watersheds, their connections, and special features. These details are the basis for describing and evaluating impacts of development activities on the physical, chemical, and biological components of natural water systems. Next we'll look at how to evaluate some specific land use activities in terms of their effects on ecosystems and watersheds—and on the quality and supply of the water that sustains us.

part II

WADING DEEPER

Identifying and Describing Impacts

3

How Land-Use Activities
Affect Water

> If sustainable development is to mean anything, such develop-
> ment must be based on appropriate understanding of the
> environment—an environment where knowledge of water
> resources is basic to virtually all endeavors.
>
> REPORT ON WATER RESOURCES ASSESSMENT, World Meteoro-
> logical Organization / UNESCO, 1991

How do our activities on land affect water resources? We consider this
question when we add a water resources protection segment to a land-
use plan, monitor a conservation easement, read a news article about water
pollution, or review an environmental impact statement (EIS). To be useful,
the answer must be specific, and this can be a challenge. It's not enough to
say that "development affects water"; we need to know which development
activity at a specific location produces an unfavorable change in specific water
resources—and how.

This chapter outlines a process for describing residential, commercial, and
agricultural development activities and their effects on water resources. The
first sections examine how we look at environmental impacts and resource
protection; subsequent sections include project description and site disturbance, construction and post-construction activities, agriculture and water
quality, and cumulative impacts.

Chapter 5 will discuss industrial and energy development and effects on
water resources at a larger scale.

Diagnosing Impacts

An activity that alters one part of an ecosystem or watershed may also affect
other interconnected parts of the same system. If you cut your foot and it
becomes infected, the infection causes a fever that gives you a headache. You
can take a painkiller to get rid of the headache, but you won't really solve the
problem until you take care of the infection in your foot. When you dump

contaminants into a stream, they are likely to travel downstream—and end up anywhere the water flows. But you can't clean up the downstream water until you have addressed the upstream source of the contamination.

Tracking the effects of land-use activities on water resources requires a diagnosis of what is harmful and why. While most of us aren't doctors, we understand the basics of a healthy body. Because of this understanding, we can often recognize when something has gone wrong within the body system. You may take an acetaminophen tablet for a headache, but if you take an overdose of the drug, it leads to liver failure and could kill you. Our bodies can withstand all kinds of stresses and remain functional, up to a point. If we exceed the ability of the body to process or get rid of a foreign substance or stressor, it will make us sick. Ecosystems work the same way. By adding contaminants that don't belong or interfering with the body's working parts, we may tip the balance to the extent that the system can no longer work properly.

It isn't possible to live on earth without creating some kind of impact. We influence the health of our water systems by dumping contaminants into the water, changing where water flows and is stored, and altering the surrounding landscape. People are part of ecosystems. We change natural communities wherever we build roads and houses, dump our trash, grow crops, or fertilize our lawns. More people means more activities that affect the environment. Over time, these activities can change natural systems to the point where they no longer provide the services we depend on.

Too Big to Fail, or Too Small to Matter?

Can our activities really change large ecosystems? The Colorado River watershed encompasses 246,000 square miles from Wyoming to Mexico. In 2013, American Rivers, a national conservation organization, designated the 1,450-mile-long Colorado as our most endangered river. Decades of river management have exploited its water for residential, agricultural, and industrial use with little regard for the river's other roles, described by Earthjustice: "the lifeblood of the west; it defines our geography, sustains our fish and wildlife, feeds and powers our cities. Without it, our lives and heritage would be fundamentally different."[1] The fate of the Colorado is the result of innumerable small decisions throughout its watershed about how to use the river's water. Like the human body, a river of this size can withstand a lot of use and abuse. However, there comes a time when uses all add up to "too much," placing the system itself in jeopardy. The human goes to the hospital; the river makes it to "endangered" status.

Chesapeake Bay is the largest estuary in the United States. Its watershed stretches across six states—sixty-four thousand square miles with more than one hundred thousand streams, rivers, and creeks. The watershed land-to-water ratio is 14:1, the largest for any coastal water body in the world. Land use within its watershed affects the water quality of the bay; more than 17 million people live in the watershed.[2] For years, we have studied the Chesapeake Bay watershed and evaluated impacts on its ecosystems. The Chesapeake Bay Foundation describes the overall health of the watershed in its *2012 State of the Bay* report: "While hopeful, a Bay health index of 32 on a scale of 1 to 100 should be a sobering reminder that there is a great deal left to do."[3]

The Mississippi River floods periodically. Hundreds of tributaries feed this 2,350-mile river that drains a watershed of over a million square miles. Draining the river valley's wetlands and altering river and tributary channels have destroyed many natural flood-protection services. Gradually, old meanders of the river were modified to accommodate millions of acres of agriculture and urbanization. In 2011, record flooding caused hundreds of millions of dollars of damage to farmlands, billions of dollars in property damage, and millions more in losses to local economies.[4]

Pollution of big water systems makes the news—when these systems fail, we hear about it. But most of us relate to water systems at a smaller scale. We look at small streams, wetlands, and ponds in our communities and figure that flooding or pollution at that small scale doesn't matter when compared to the dramatic results of big systems gone awry. If you live on a half acre in the Chesapeake Bay watershed, how could your small-scale activities possibly affect that system?

Myriad subbasins make up these large watershed systems. At a small scale, decisions by local leaders may influence only one small subbasin or building project, but the impacts from many small decisions accumulate across a large area. By the time water quality diminishes to the extent that it attracts regional attention, the task of cleaning up a big watershed or river valley is formidable and expensive.

What about Laws That Protect the Environment?

Don't we have local, state, and federal regulations in place to prevent degradation of our water? Why should anyone outside the regulatory community examine environmental impacts? Regulations are an important tool for natural-resource protection, but they do have limitations. Look at the examples; despite the regulations in place today, the Colorado River, Chesapeake

Bay, and Mississippi River systems—and their watersheds—have not been adequately protected.

Many of us would expect to begin with regulatory requirements as a basis for evaluation. But this strategy is like "teaching to the test"; it doesn't present the whole story, but meets only the letter of the law. Such a review is likely limited by scope, definition, and regulatory interpretation. Federal and state government agencies have not updated many laws to reflect the latest scientific research about watersheds and ecosystems and how they work—and their benefits to human communities. Environmental protection laws vary among states and municipalities; resources that are protected in one county or state may not be protected in another, even when regional governments share the same watershed, river, or lake. Across the country, the level of environmental analysis required by federal, state, and local regulations is often inadequate to protect our natural water systems effectively.

The competition for water is increasing. Different interests compete for a river's water—drinking water for a downstream city, aquatic habitat to sustain a fishery, recreational waters for canoeing or rafting, or irrigation for crops. Every planned development project can affect water quality and supply. The problem is, we don't always pay attention to the details of these changes and whether the result is beneficial or harmful.

How You Look at It

Different goals for land use lead to different views about the consequences of projects and activities. Sometimes the review process jumbles together the description of existing conditions, negative impacts, and mitigation. Those who support a particular project may assume it won't have significant negative effects, or that once we can no longer see a problem, it's gone—like the globs of oil on the water surface in the Gulf of Mexico after the 2012 BP Deepwater Horizon oil leak. Treated with the chemical dispersant Corexit, the oil sank below the surface, and when it could no longer be seen, many people assumed it disappeared and was no longer a problem. But the oil is still there, made even more toxic in combination with the Corexit.

Different assumptions about the value of natural resources influence the way you look at a site or project. It's important to protect natural resources, and it may also be important to build the project. Two different ways of looking at the same piece of land yield two different sets of conclusions about potential environmental impacts. This variety of perspectives improves project review. A local planning board conducted a site visit with a developer proposing a residential subdivision. A volunteer representative of the local conservation commission was part of the site-review team. Later, when the full committee reviewed the site plans and comments, she lamented that she hadn't said much on-site, even though she noticed some obvious problems concerning wetlands, runoff, and drainage patterns. "I didn't have the expertise to comment!" she said. "I was out there with all these big experts and figured they would see it all. When they didn't say anything, I figured I must be mistaken." Yet she had correctly identified some issues that no one else had addressed. While expert reviews are essential, experts too may be constrained by "viewpoint," so we need to evaluate all perspectives when looking at a site.

Are natural resources assets to the community, or are they constraints to development? If we consider them only as constraints, we devalue them as a cost rather than a benefit. How much water pollution will we tolerate? Is a stream or a wetland a liability or an asset for a small town? Sometimes project design doesn't take into account a site's natural features or assets. The value of water resources extends beyond their ability to move water from one place to another. Our understanding of ecosystems and watersheds affects how we describe the activities that can change them.

Environmental Impacts: "So What?"

We may report that a project will require the clearing of fifty acres of forested land. But that is just the beginning—the consequences from that clearing are like the magician's scarves hidden in a hat: You pull one out and find it attached to another scarf, and yet another. And then two or three at one time, each knotted to its own rope of scarves. To get a handle on these connections, you might ask, "So what?" ("Why is that important?") after each "scarf" (project activity or environmental change).

If you clear that fifty acres of forest, you expose the soil to increased erosion via stormwater runoff. Residential development that creates more impervious surface will replace the forest and its ability to protect water quality and aquifer recharge. So what? Well, when you increase impervious surface, less water soaks into the ground where it falls, and runoff increases. So what? Less water is available to replenish groundwater. In addition, increased runoff carries a larger contaminant load, which winds up in a nearby stream. So what? One of these contaminants is road salt from deicing, and its presence in the stream changes water chemistry. So what? Some of the fish that live in that stream will die, and fishing will no longer be a recreational asset to the community.

You can follow this type of progression for each activity listed in the tables that follow. For every activity there are many connected "scarves"—possible effects depending on site, project, and ecosystem.

We can measure the concentration of a pollutant like road salt in a stream, but until we describe what that means to aquatic life and ecosystem benefits (for example, fishing and the presence of potable water in our wells), we haven't really described the full effect of that contamination. Sometimes we don't notice small changes until they have become extreme enough to catch our attention and inspire action.

People do notice when the benefits are gone. A town in a rapidly developing suburban area required a wetland boundary verification to meet the requirements of a local wetland-protection law. The site was one of the last bits of open land in the area—3.5 acres sandwiched between residential development on one side of the highway and shopping mall parking lots on the other. The wetlands were the area's last remnants, muddy with silt and overcome by invasive plants. The silt fencing put in place to protect the wetlands from the sediment load in stormwater runoff straggled along the edge of the water; bulging pockets of sediment sagged into the mud. Culverts discharged streams of muddy water into the wetland at both ends. As I mucked along the water's edge, a man from one of the neighboring houses approached to ask what I was doing. His comments were unusual—he wanted to talk about the spring peepers. "We don't hear them anymore," he commented, shaking his head and looking at the wetland. "Look at this. Peepers used to be so loud this time of year. They were everywhere. But now in our neighborhood it's so quiet. They're all gone." Sadness, loss, regret. Fields of parking lots, trash, dirty water spewed by culverts and pipes, habitat gone, and no place left for peepers.

The community might have avoided this outcome if the municipality had identified wetland habitats that were of value to the community before they were lost; impacts from surrounding construction activities on those habitats; and a mitigation plan to protect the peepers' habitat.

Project Description and Site Disturbance

When you review a project, keep in mind point of view and interconnected water resources. Look at the project site and the maps you compiled as described in chapter 2. Your review of natural resources should include adjacent areas where habitat and water systems extend beyond project site boundaries. Consider the total size of the stream or wetland, on-site and off-site. A project's effects aren't confined to the water that lies within the project site boundaries.

Next, describe the project, including all phases over time, from start to finish. The account of project activities should include the following:

- Number of lots / dwelling units
- Extent of roads (length and width)
- Utilities and other infrastructure
- Stormwater management facilities
- Extent of all other impervious surfaces (parking lots, maintenance or storage facilities, swimming pools, and outbuildings)
- Source of water and estimated water use per household
- Wastewater treatment (sewage treatment or individual septic systems)

You can outline on a map the total area to be disturbed by the project. Include the entire area from which existing vegetation will be removed, and all areas where soil will be compacted (all roads and other surfaces where trucks or heavy equipment will be operated or stored). Some of these areas may be replanted after construction, but they are still considered part of the area of disturbance, because it may not be possible to restore soils to pre-disturbance condition, and new vegetation may not replace lost habitat. Water resources enlarge the "area of disturbance" beyond site boundaries as water moves on and off the site, above and below the ground, and downstream.

Analysis of the project and its site tells us the extent to which land and water resources are likely to be physically or chemically changed, and from that point we can describe additional "so what" impacts.

The following text and tables describe typical activities associated with residential development. The list of activities is not all-inclusive or intended as a replacement for detailed on-site evaluation by experts. You may need to hire the appropriate professional(s) to analyze wetlands, water quality, ecosystems, or hydrology, and to evaluate specific consequences (in detail) from various project activities. This process will help you organize information so you can look at the effects of individual activities and combine them for an assessment of cumulative impacts.

Project Activities

Project activities fall into two broad categories. "Construction" includes all activities associated with preparing a site, storing materials, constructing roads and buildings, and installing infrastructure. "Post-construction" encompasses the use of the land after construction has been completed and includes water use and waste treatment.

The degree to which these activities (individually or collectively) affect specific natural resources depends on the following factors:

- The type of activity, how large it is, and how long it lasts (for example, doubling the extent of impervious surfaces within a small subbasin)
- The time of year when the activity occurs in relationship to seasonal water or biological cycles (for example, stream-crossing construction that requires streamflow to be diverted during fish spawning season)
- The location of the activity in relationship to sensitive or significant natural features (for example, the construction of a road through a turtle-nesting area)
- The sensitivity of the site's natural-resource features or systems to the specific changes associated with each activity (for example, the presence of mayfly nymphs, which require high-quality water)
- The cumulative impacts that have the potential to change or disrupt ecosystems or wetland functions and services (for example, discharges from multiple septic systems at individual cottages surrounding a small lake)

The tables in this chapter present an overview of construction activities and examples of some of their consequences. Each of the activities listed may or may not apply to a specific project, depending on project and natural-resource details. Activities may produce additional consequences depending on regional, site, and natural-resource characteristics. Environmental impacts may be major or minor, and some can be mitigated, while others cannot. (Mitigation will be discussed in more detail later.)

Site Preparation and Construction

Use the activities listed in table 3.1 as a basis for the "so what" exercise as you explore their consequences and connections. Additional details about these activities will sharpen your understanding of how project activities are likely to change specific components of ecosystems or watersheds.

Clearing Vegetation

Identify the vegetation and its location in relationship to water resources (buffer, floodplain, headwaters, etc.). What species will be removed, and are they rare, protected, native, or invasive? How large is the area to be cleared? What is the role of this vegetation in the ecosystem or watershed?

TABLE 3.1. Overview of site preparation, construction activities, and examples of impacts

Site preparation and construction	Examples of potential impacts
Clearing vegetation	Loss of species and habitat; habitat fragmentation; increased soil erosion, runoff volume, and sediment load; decreased groundwater recharge
Grading	Changes in drainage patterns and water supply to streams, ponds, wetlands; soil compaction; decreased groundwater recharge; increased erosion
Filling or depositing materials in wetland or stream	Habitat loss; degraded water quality; increased flood potential
Draining or dewatering	Change in watershed services; altered habitat; loss of spawning or breeding habitat
Stream channel deepening or straightening	Increased downstream flood potential; water quality degradation; habitat loss; increased erosion
Water crossings and culverts	Bank disturbance; increased erosion; changes in water-flow patterns; interrupted wildlife movement patterns; increased flooding; degraded water quality
Construction of roads, streets, buildings, infrastructure	Soil compaction; increased impervious surfaces; increased erosion, runoff, and pollutant load; changes in drainage patterns; habitat fragmentation or loss
Disposal of construction materials and wastes	Water quality degradation (surface and groundwater); habitat changes
Movement and storage of heavy equipment	Soil compaction; loss of vegetation; increased stormwater runoff; decreased groundwater recharge; floodplain disturbance; water contamination; habitat loss
Stormwater management infrastructure (engineered practices)	Concentrated contaminant load in basins and outflows; changed water supply to wetlands, lakes, and streams; decreased recharge; increased potential downstream flooding
Erosion control practices (improperly sited or not maintained)	Increased runoff, sediment load, erosion
Changes in land use: converting open land to developed land	Altered hydrology and drainage patterns; loss of watershed and ecosystem services; water pollution; increased stormwater runoff; decreased groundwater recharge; increased flooding potential

The plants that are removed may include rare species, or other natives that are hard to replace. Mature trees take years to grow to a size that enables them to play a valuable role in the ecosystem. The extent of vegetation removed may change a watershed's ability to perform important services; the location of the clearing may change ecosystem functions and the role of the plant community—providing buffer and bank protection; furnishing habitat and food for wildlife; holding soil in place; slowing runoff or floodwater; and facilitating groundwater recharge. Clearing vegetation along the water's edge or within a stream, lake, or wetland buffer changes habitat and may diminish buffer services such as water temperature moderation; nutrient input (from plant leaves, stems, twigs); food-chain support for aquatic or wetland organisms; and protection from adjacent land-use noise, light, and physical disturbance. Clearing vegetation within a wetland alters habitat and hydrology and diminishes many services that the wetland provides, such as the ability to improve water quality.

Each area cleared of vegetation also creates an "edge effect," which extends into adjacent intact habitat and influences plants and animals beyond the cleared area. For example, the effects of clearing a patch of forest extend approximately three hundred feet beyond the cleared edge. Within this edge, conditions may favor invasive plants and push sensitive species back into the intact forest.

Clearing vegetation fragments habitat, dividing large patches into smaller areas that may not support the same mix of species. Fragmentation also occurs when barriers to wildlife movement are created, like roads or barren strips that make it difficult for species to move between nesting and feeding areas. Fragmented habitat generally favors species that are adapted to disturbed conditions and discourages species that require high-quality habitat and larger habitat areas.

The timing of vegetation clearing may affect species that depend on a particular habitat during different life-cycle stages—for example, birds nesting in streamside shrubs, or bats roosting in trees during summer. In these cases habitat is lost when trees or shrubs are removed, no matter the season of cutting—removal during winter may not kill nestlings or bats outright but results in loss of habitat for the following summer. Different plant communities provide characteristic services, so it's important to identify them as specifically as possible.

Vegetation clearing leaves the soil exposed and vulnerable to erosion, increasing the sediment load in stormwater runoff and the potential contamination of nearby surface waters. Exposed soil is also vulnerable to colonization by invasive species, which tend to dominate disturbed areas, preventing revegetation by native species and decreasing the site's habitat value.

Grading

Site grading changes the way water moves across the land and alters where it is stored and where it soaks into the ground. Heavy equipment compacts soil, eliminating the pores between soil particles that hold air or water, decreasing groundwater recharge and increasing stormwater runoff. Grading within a buffer changes its ability to protect the water quality and habitat of the adjacent wetland, stream, or lake. Site grading may increase or decrease the water supply to individual wetlands, lakes, and streams—altering the distribution of water in the watershed, composition of plant communities, and water available for aquatic habitats. Grading changes wetland hydroperiod. This in turn may cause ponding or inundation that changes wetland plants and functions, such as removal of water contaminants by particular plant species. In addition, grading exposes soil to erosion and increases likely stormwater runoff into surface waters—affecting water quality, habitat, flood protection, and other system functions.

Changes in drainage patterns on steep slopes lead to increased erosion and a higher sediment load in stormwater runoff. Does bulldozing an existing steep slope solve the problem or create a new one? During a site-plan review for a residential / golf course development, residents in a rural Hudson Valley town expressed concern about impacts from grading and construction on steep forested slopes. The developer was pleased to respond that there wouldn't be any effect on steep slopes—because he was planning to remove and grade slopes into a level area that could accommodate more houses. The planning team challenged the grading plan and required an accounting of all proposed changes to the original drainage patterns on the site. As a result, the plans were amended to avoid most of the site's steep slopes.

Grading may disturb former agricultural (or golf course) soils that have a high residue of pesticides or other chemicals, moving contaminated soil to the surface where stormwater runoff can transport it into surface waters.

Filling or Depositing Materials in Wetlands or Streams

Depositing fill and other construction-related materials into a small wetland or stream can reduce the ability of these ecosystems to store water (such as runoff and floodwaters). The partial filling of a wetland or stream can also change bottom habitat, introduce contaminants to the water depending on the characteristics of the fill, change water depth, and alter stream flow or wetland hydroperiod. These changes in turn can influence the type of plants that are able to grow and the habitat or water-quality improvement services they provide.

Draining or Dewatering

These activities change the hydrology of a wetland, lake, or stream and the ability of these systems to function—for example, to slow and store floodwaters, support habitat and species, cycle nutrients, and replenish groundwater supplies. Dewatering can destroy small wetland and stream ecosystems by killing plants and eliminating aquatic habitat. Depending on the season, temporary dewatering of stream channels can interfere with fish and amphibian spawning or breeding cycles. Heavy-equipment operation within a lake, wetland, or stream compacts the bottom sediment, destroys habitat for bottom-dwelling organisms, and contaminates water with oil, grease, fuels, and other substances associated with equipment operations.

Stream Channel Deepening or Straightening

Healthy streams typically have a meandering pattern, which slows water flow, and areas of floodplain where floodwaters spread out, slow down, and gradually seep into the ground. Activities that widen, deepen, or straighten stream channels to send water quickly off-site can increase bank erosion, increase water depth, and cause water to move faster, which in turn changes aquatic and stream-bottom habitat. Channel straightening and deepening can direct water away from floodplains and their flood-control services. Dredging a stream for flood protection can create conditions that actually increase flood potential downstream. There is often no single remedy for reducing flooding—it should be treated as part of an overall watershed management strategy that includes protecting wetlands and small streams; regulating construction in floodplains; rehabilitating impaired banks and stream channels; reducing sediment deposition and erosion; and integrating structural changes such as bioengineering in areas most vulnerable to flooding.

Water Crossings and Culverts

Construction of bridges and other water crossings may require removal of trees and shrubs needed for bank stabilization—resulting in increased bank erosion and changes to water quality, depth, and flow patterns. Bridge and culvert design should address seasonal dewatering as well as climate-change-induced effects such as increased stormwater runoff and flooding.

Culverts quickly convey water; depending on their size, type, and location, they can direct water away from one wetland or stream and increase flow into another. They can change water supply and depth and disrupt seasonal water-level fluctuations (hydroperiod). Where roads or driveways bisect wetlands, installation of a single culvert may not permit sufficient water exchange from one side of the wetland to the other. Flow constrictions cause upstream ponding in wetlands, which in turn may lead to increased tree mortality and sediment accumulation. Undersize or otherwise inadequate culverts may cut off wetlands or streams from a portion of their water source. Installing a series of culverts may address this problem in some situations.

Culverts do not improve water quality, nor do they reduce flood flows. Contaminated water that enters one end of a culvert is just as contaminated when it flows out the other end. In some situations, culverts that collect runoff from a variety of locations may concentrate stormwater contaminants, directing them into a single outfall area or discharging into a stream, lake, or wetland. All water discharged onto the land via culverts should be spread as sheet flow across vegetated buffers before it enters any wetland, lake, or stream. Unless the culvert is an open-bottom box design, it separates the water it carries from other ecosystem features—for example, it won't allow for water seepage into the ground or support habitat for stream-bottom organisms.

Culverts require regular cleaning and maintenance or they fill with sediment over time; sedimentation reduces their capacity to carry water flows generated by storms, sustain wetlands and streams, and allow critical seasonal water-level fluctuations.

Construction of Roads, Streets, Buildings, Infrastructure

Construction usually increases impervious surfaces, which can affect water resources by

- increasing stormwater runoff and its contaminant load;
- changing the amount of water that flows into specific streams, wetlands, and lakes;
- decreasing groundwater recharge;
- increasing flooding;
- eliminating headwater streams or small wetlands; and
- fragmenting or degrading habitat.

Construction along a stream's edge separates the channel from its floodplain and reduces the system's ability to retain and absorb floodwaters.

Depending on location, buildings and roads fragment habitats and create barriers to wildlife movement between habitats—for example, preventing turtles from moving between pond and gravel-bank nesting sites, or salamanders moving between forest and breeding pools. Standard curbing along roads and residential streets is a barrier to the migration of small animals such as salamanders, frogs, and turtles.

Roadside ditches, if not constructed properly, may convey stormwater and its sediment load away from some water resources and redirect it into other areas. Ditches can change wetland and stream hydrology. If the ditches are V-shaped with exposed soil, they have greater potential for erosion and increased stormwater sediment loads than do ditches that are U-shaped and reinforced with rock or gravel.

Disposal of Construction Materials and Wastes

Stormwater runoff may carry improperly contained construction materials and wastes (such as paints, adhesives, preservatives, and other chemicals) into surface waters, or polluted runoff may seep into soil to contaminate groundwater. Construction scraps may contain chemicals that contaminate water and soil (such as arsenic from pressure-treated wood, formaldehyde from composite wood products, fiberglass insulation, and melamine resin). Wastes deposited directly into wetlands, lakes, or streams may impede water flow, destroy habitat, reduce flood storage capacity, and degrade water quality.

Movement and Storage of Heavy Equipment

The movement or storage of heavy equipment compacts soils, compressing soil layers and eliminating air and water between soil particles. Soil compaction reduces the soil's ability to absorb water, recharge groundwater, and support soil life (including plant roots). Heavy-equipment operation can damage vegetation and habitat; increase stormwater runoff; disturb banks and floodplains; damage ecosystems; and contaminate soil and water from fuels, oil, grease, and other contaminants associated with operation.

Stormwater Management Infrastructure

Highly engineered stormwater management practices concentrate contaminant loads in basins and outflows; change water supply to wetlands, lakes, and streams; decrease groundwater recharge within a subbasin; and may increase potential downstream flooding. Green infrastructure, best management practices, and low-impact design are preferable alternatives for managing stormwater runoff. Whenever possible, project designs should preserve a site's natural stormwater management capability and avoid highly engineered practices. Best-management practices preserve natural drainage features and allow precipitation to seep into the ground as close as possible to where it falls. Constructed stormwater management facilities are variably effective depending on (1) how well they are designed to match site conditions, (2) whether they are constructed exactly as designed, and (3) whether they are maintained regularly. If they aren't maintained, they can contaminate water resources, disrupt aquatic ecosystems, and fail during severe flood events. These potential failings are even more significant in light of climate change and a higher frequency of severe storms. More highly engineered stormwater management infrastructure requires more maintenance to sustain its effectiveness. Clear identification of a reliable entity responsible for long-term maintenance improves the chance that stormwater facilities will remain effective into the future. Use of specific Low Impact Design and Green Infrastructure practices, as described by the Environmental Protection Agency, improves water resource protection. The preferred alternative is to leave natural stormwater management systems intact wherever possible—natural stormwater management (such as intact floodplains, protected wetlands and buffers) costs less and causes fewer impacts on ecosystems or natural watershed functions.

The goals of stormwater management don't always match the goals of protecting ecosystems or natural watershed functions. Some stormwater management practices and facilities negatively affect these systems, especially when they aren't designed to protect specific habitats, or aren't regularly maintained for optimum functioning. For example, detention ponds do not remove all contaminants from stormwater runoff but typically are designed to comply with regulatory mandates and "minimum" pollutant-removal standards. When these ponds or basins are constructed too close (less than one hundred feet) to wetlands, ponds, and streams they can concentrate water contaminants from stormwater runoff into a small area and direct them into surfaces waters via outflow culverts. They also can provide "false" habitat that may attract species for breeding but that impairs egg and larvae development because of contaminant loads (such as pesticides or road salt).

Stormwater management can change natural drainage patterns—depriving aquatic and wetland habitats of an adequate water supply; changing seasonal hydrology, hydroperiod, and stream flow; and flooding wetlands by redirecting runoff into them. Long-term wetland flooding increases water depth and can significantly change plant composition, water chemistry, and wetland functions (such as the ability to remove contaminants).

The preparation of a stormwater pollution prevention plan (SWPPP) is not a guarantee that all of a project's effects on water quality and runoff management are resolved.

Erosion-Control Practices

Silt fencing is commonly used to retain sediment loads carried by stormwater runoff and keep them from washing into surface waters. Its effectiveness depends on how appropriately it is placed and how well it is maintained. When improperly maintained, the fencing accumulates sediment—bulging, sagging, and eventually failing to contain stormwater flows and sediment. You can improve water-quality protection by installing a double row of silt fencing along the outer edge of a vegetated buffer surrounding all water resources. However, it must be sited so that it doesn't pose a barrier to seasonal migration of small animals such as turtles or pool-breeding amphibians.

Natural vegetation is an efficient and cost-effective solution to erosion prevention and control. By minimizing areas of disturbance and promptly revegetating disturbed areas with native plants, you can reduce and contain erosion from disturbed land.

Changes in Land Use

Converting land from forest or field to impervious surfaces, and from native vegetation to turf grass or nonnative landscaping, changes watersheds and ecosystems and their benefits on a large scale. The magnitude of these changes varies depending on natural features, type of development activities, surrounding land use, and cumulative impacts. Development generally increases impervious surface within a watershed; reduces land area for groundwater replenishment; removes habitat, corridors, or connections; changes vegetation; and increases contaminant levels in stormwater runoff. Appropriate planning and mitigation can avoid or mitigate many of these impacts.

Post-Construction Activities

After construction, a different set of activities may affect a site's water resources. Table 3.2 summarizes these activities and some examples of their consequences.

Future Site Disturbance

Lot lines that extend into a stream, wetland, or lake and their buffers may impinge on sensitive areas. Future activities, like the construction of swimming

TABLE 3.2. Overview of post-construction activities and examples of impacts

Post-construction residential activity	Examples of potential impacts
Future site disturbance	Increased impervious surface; water contamination; habitat loss
Household water use	Lowered groundwater and aquifer levels (from increased pumping) and reduced water supply to wetlands, streams, and lakes
Household wastewater disposal	Increased water temperature and altered water chemistry from wastewater treatment effluent; contamination of rivers and streams from combined sewer system overflows
Septic systems	Contamination of groundwater (from old or improperly maintained septic tanks); altered water chemistry; degraded aquatic habitat; human health impacts (drinking-water contamination)
Pesticides, herbicides, fungicides	Chemical contamination of surface waters, groundwater, and soil; harmful effects on wildlife and human health; reduced populations of native insects, including bees
Fertilizers	Algae blooms; depletion of dissolved oxygen in water; harmful effects on aquatic organisms and ecosystems
Increase in turf (lawns)	Chemical contamination of stormwater runoff (carries chemicals from herbicides, pesticides, and fertilizers into water); altered water chemistry; habitat loss; toxic effect on aquatic organisms; decreased biodiversity
Landscaping with nonnative species	Reduced species diversity; altered or lost habitat; reduced native insects as base of food chains; decreased bird diversity
Disposal of yard (or other) wastes	Changed water chemistry and bottom substrate; depletion of oxygen in water; degraded aquatic and edge habitat
Disposal of pet wastes	Water contamination (via stormwater runoff) due to bacteria and parasites; harmful effects on human and wildlife health
Pavement deicing and snow removal (road salt)	Loss of vegetation (from use of road salt); harm to aquatic organisms; drinking-water contamination; altered water chemistry; degraded aquatic habitat; increased invasive plants
Increased vehicular traffic	Pollution from vehicle emissions that make their way into water
Off-road vehicles	Disturbance of wetland and stream ecosystems; damaged vegetation and habitat; soil compaction and increased erosion or runoff

pools or mowing to the edge of the water, can degrade water quality, damage buffer or banks, and destroy habitat. Once lot lines are set, you have little or no control over what happens on individual properties in the future; whenever possible, new lot lines should exclude sensitive water resources or their buffers. Actions that take future land-use activities into account (local zoning codes, local water resource protection laws, building permit reviews) can help to protect local water resources.

Household Water Use

Residential developments use water. The average family of four uses about four hundred gallons of water per day. Whether this water comes from individual or municipal wells or from surface waters like lakes and rivers, residential use can cumulatively affect water sources—for example, by lowering the water table or lake level, reducing river flow, and changing the direction of groundwater movement. During a drought or in dry climates, well pumping and water use may significantly deplete local streams, wetlands, and underground water storage. Some local governments have passed laws that require low-water-use faucets, toilets, and showers in new buildings or renovations. Depending on the regional availability of water and the health of local water sources, increased water demand can reduce the reliable supply of water for future human use, as well as change the hydrology of water-based ecosystems, influencing water quality, habitat, and other system services.

Wastewater Disposal

Household wastewater contains an array of chemicals from personal-care products, household cleaners, and pharmaceuticals. These chemicals make their way into groundwater via individual septic systems or from surface water in the outflow from wastewater treatment plants; the treatment plants kill bacteria and rid the water of nutrients and solids but are generally not designed to remove all chemicals and other contaminants. Water-softening systems and road salt introduce salt into wastewater, altering water chemistry and affecting aquatic organisms. Wastewater outflow, depending on its chemical content, may raise water temperature, increase stream flow, change water chemistry and habitat conditions, and harm aquatic organisms. These effects are more pronounced when low water levels reduce a stream's capability to dilute effluents.

Combined sewers are older systems currently used by hundreds of cities in the Northeast, Great Lakes, and Pacific Northwest regions. They collect stormwater runoff, domestic sewage, and industrial wastewater in the same pipe-and-culvert system, transporting all to a sewage treatment plant, where water is treated and then discharged into a river or stream. However, heavy rainfall, leading to increased stormwater runoff, can overload the treatment plant's collection system and exceed its capacity. As a result these systems overflow and discharge excess wastewater (including raw sewage) directly into streams or rivers. The wastewater carries elevated levels of bacteria (fecal coliforms) that can contaminate a river or stream to the extent that it becomes a human health hazard. Municipalities across the country that use combined sewer overflow systems experience periodic contamination in local waters after heavy rainfall; it is a major water-pollution problem.

Septic Systems

Septic systems can contaminate surface and groundwater when they are installed in unsuitable locations (for example, unmapped wetlands, floodplains, or soil with a seasonal high water table); are poorly maintained as they age; or are installed at a density that exceeds the soil's capacity to process their effluent.

Local health department procedures, requirements, and enforcement vary; check local guidelines to determine whether they provide adequate protection for water resources.

A recent study from Dutchess County, New York,[5] describes one method for guiding residential development so that it doesn't exceed the natural environment's capacity to support individual septic systems and replenish aquifer supply. This approach uses information about local rainfall, aquifer recharge rates, and soils to develop limitations on the density of residential septic systems that a given area can support. Suitable parcel size is related to soil characteristics—areas with well-drained soils, high precipitation, and high aquifer recharge rates can sustain denser residential development. Larger parcels (and lower residential density) are appropriate in areas where precipitation and aquifer recharge rates are lower and soils are poorly drained. Development at a density that exceeds the soil's capacity to process human wastes risks eventual groundwater and well contamination. Flooding intensifies the risks of surface-water contamination from improperly sited or aging septic systems.

Pesticides, Herbicides, Insecticides, Fungicides

Commonly used herbicides, fungicides, and insecticides contain an array of chemicals that are often harmful to "nontarget" organisms or to human health—even when used according to manufacturers' directions. Landscaping, lawns, and street or infrastructure rights-of-way typically rely on periodic applications of these chemical products. We are so accustomed to their use that we may forget that many of the chemicals they contain have not been thoroughly tested for ecosystem effects or cumulative impacts on the environment. Carried by stormwater runoff into nearby wetlands, lakes, or streams, many of these chemicals are harmful to aquatic organisms, water resource ecosystems, and human health. Depending on site conditions, some of these chemicals may also find their way into the water table and on-site wells.

Certain pesticides harm beneficial species. For example, bees pollinate many food and garden plants. In the United States, pollination contributes to $20 billion to $30 billion in agricultural production annually; honeybees pollinate a variety of crops including almonds, avocados, oranges, apples, and cranberries. Bumblebees pollinate many fruit and berry crops. A growing body of peer-reviewed information tells us that pesticides containing neonicotinoids (neonics), a class of insecticides, are most harmful to bees. Even small doses may lead to flight and navigation problems, fertility issues, and inefficient foraging—making colonies more susceptible to illness or parasites. These pesticides persist long-term in soil and water and are also toxic to soil-dwelling insects, benthic aquatic insects, and grain-eating vertebrates. If we kill bees, we risk significant harm to ecosystems as well as crops and gardens. Research from the USGS looks at the effects of neonics on surface waters and documents widespread presence in streams and rivers in the Midwest and aquatic toxicity levels.[6] To avoid this damage, we should not use some pesticides (such as neonicotinoids) at all.

Pesticide chemicals may enter our water via stormwater runoff or the air. We should avoid all pesticide use within surface waters and their buffers. In some cases, we make exceptions for the control of invasive or other harmful species, weighing cost to the environment against cost of harm from these species and effectiveness of alternative, nontoxic means of pest control. Alternatives include biopesticides, made from natural plant, animal, or mineral materials; integrated pest management, which includes biological control of pest species; and other available control methods that produce the least harm to the environment.

Fertilizers

Stormwater runoff carries fertilizer used on agricultural crops, landscaping, lawns, and gardens into surface waters. Fertilizer nutrients—notably nitrogen and phosphorus—contaminate the water, accumulate to levels that aquatic or wetland ecosystems cannot process, and lead to algae blooms. As the algae die, the decomposition process removes dissolved oxygen from the water. Lack of sufficient oxygen in the water kills fish and other aquatic organisms. The smaller the wetland, stream, or lake, the more easily excess fertilizer can contaminate it. The cumulative effect from multiple homes can be significant where vegetated buffers are not in place to remove fertilizer from runoff before it enters surface waters.

Lawns

The replacement of native vegetation and habitats with lawns produces a significant portion of the chemical contamination (herbicides, insecticides, fertilizers) of water resources in residential areas. Lawns increase the volume of stormwater runoff (they aren't as effective at absorbing and retaining water as native vegetation, especially during heavy rains). Turf grasses provide little or no habitat value, replacing other habitats that support higher species diversity. In addition, lawns require a high level of maintenance and regular use of fertilizers, herbicides, and pesticides.

The chemicals we apply to lawns often end up in our water. When lawns replace the buffer vegetation along the edges of lakes, streams, or wetlands, these chemicals are more likely to find their way into the water. An estimated 40 million acres of lawn in the United States[7] receive about five to seven pounds of pesticides per acre per year.[8] Homeowners use three million tons of fertilizer annually on lawns, according to a study from Duke University; nitrogen fertilizer use could be reduced by 50 percent if grass clippings were left on lawns.[9]

Landscaping with Nonnative Species

The replacement of native vegetation with nonnative species seems harmless when you look at one backyard at a time. But it has far-reaching consequences when you consider multiple backyards—it influences which species of plants and animals will be able to thrive in and near residential developments. Areas planted solely in nonnative species—even if these shrubs provide physical cover and nest sites—don't provide the food sources and full range of services needed to sustain local biodiversity. For example, native insects often cannot fully digest nonnative plant leaves and stems; if these insects cannot thrive, they are not available as a critical source of protein for nestling birds. Loss of these insects can reduce birds' reproductive success.

Landscapers often introduce invasive plants like Japanese barberry and Norway maple as nursery stock. New development and landscaping projects should use native plant species whenever possible, and in consultation with wildlife specialists or biologists who can guide revegetation efforts to produce sustainable native habitat. Introduction of nonnative plants into any stream, lake, or wetland changes habitat and can crowd out native species and disrupt ecosystem processes.

Disposal of Yard Waste

Throwing yard waste (such as brush cuttings, grass clippings, and wood chips) into a nearby wetland or stream is a common practice. Streams carry the dead branches and other waste out of sight. But as these "natural" wastes decay in the water, they deplete dissolved oxygen and change the water's chemistry. If the volume of waste is too large, it may exceed the capacity of a small stream or wetland to process it, removing oxygen and weakening or killing aquatic organisms. Yard debris can also catch on the stream bottom and banks, changing water flows, reducing the stream's capacity to hold stormwater, and increasing flood potential.

Disposal of Pet Wastes

The American Pet Products Association estimates that 47 percent of US households own at least one dog;[10] some seventy-eight million dogs in the United States generate 10.6 million tons of waste per year.[11] Large residential areas with many pets generate wastes that can be a significant water-quality problem. For example, in urban areas 60 percent of dog waste is thrown into the trash, and 40 percent is left on the ground, often in concentrated areas close to human activity. When owners dispose of pet waste improperly, stormwater runoff washes it into storm drains (which may or may not be connected to treatment facilities) or directly into nearby surface waters. Untreated animal feces can significantly contaminate lakes, wetlands, and streams; they contain nutrients that promote weed and algae growth, leading to decreased oxygen in the water (eutrophication). These wastes also contain viruses, bacteria, and parasites that can infect aquatic organisms, wildlife, and people. One study from Seattle, Washington, shows that nearly 20 percent of the bacteria found in area watersheds came from dog waste.[12] EPA provides guidelines for proper disposal of pet wastes to protect our water.[13]

Pavement Deicing and Snow Removal

In cold climates, a variety of deicing chemicals and sand mixtures applied to paved surfaces deliver a significant load of salt to our wetlands and streams. "Road salt" (sodium chloride) washed from roads into stormwater drains or ditches, or directly into surface waters, changes water chemistry and can harm aquatic organisms. An elevated concentration of salt changes the type of plants that will thrive in a particular wetland, stream, or lake ecosystem, favoring salt-tolerant species, and can degrade the entire system. Salt that enters groundwater and wells can ruin drinking-water supplies. Road salt and other contaminants are often concentrated in snow scraped from road surfaces by plows. When plows push piles of snow into wetlands or stormwater basins, the salt and other contaminants accumulate and become more concentrated as the snow melts and evaporates. Evaluate the use of alternative road deicers to compare potential effects on water resources. Area-specific management plans can reduce the amount of deicers that wash into sensitive water resources.

Increased Vehicular Traffic

Gas-powered vehicles (cars, trucks, and heavy equipment) on local streets and roads release pollutants into the air and onto road surfaces; many of these contaminants find their way into our water.

Off-Road Vehicles

The extensive use of off-road vehicles within wetlands or small streams can damage these ecosystems by compacting soils, destroying vegetation, degrading habitat, and changing water flow or storage capacity. Use of such vehicles should be restricted to designated pathways that avoid wetlands and streams.

Agriculture and Water Quality

Residential areas aren't the main source of chemical and fertilizer pollution; 80 percent of all US pesticide use is in agriculture. Herbicides are the most widely used in both agriculture and residential areas.[14] The chemicals in these products vary in their toxicity to nontarget organisms and ecosystems, effects on human health, and persistence in the environment. Some of them break down into harmless substances over time; some persist long-term. Others make their way into food chains or accumulate in the soil until disturbance brings them back to the surface, where stormwater runoff washes them into streams and wetlands. Many agricultural insecticides, herbicides, and fungicides contain chemicals that can severely harm nontarget plants and animals, natural systems, and human health. Specific consequences vary depending on the chemical constituents of each product, the concentration in the environment, and the sensitivity of ecosystems and organisms.

Agricultural use of fertilizers causes the same water-quality problems as residential use, but on a much larger scale—and with a correspondingly larger potential to degrade water resources. When improperly contained, livestock and poultry waste and its load of nutrients washes into our water, causing significant contamination. Antimicrobials and hormones fed to livestock are present in manure and enter water systems via runoff or seepage. These chemicals pose risks to human health and the environment; continuing research documents the details.

The cumulative effect of widespread chemical and fertilizer use is water-quality degradation on an increasingly large scale. While this degradation isn't entirely from agricultural practices, agriculture is a significant source of water-quality degradation within watersheds. Industrial agriculture is particularly culpable because of the scale of its operations. But even on a small scale, fertilizers and other agricultural chemicals in runoff wash into small streams within subbasins; the load increases as tributary streams flow into larger rivers and contaminants are carried along. Individual contaminants may be deposited in sediment or soil, or accumulate in ponds, wetlands, or basins that collect stormwater.

Cumulative Impacts: The Multiplier Effect

A cumulative impact is the result of individually minor but collectively significant actions taking place over a period of time. Multiple residential, commercial, and agricultural activities contribute to cumulative impacts on natural systems. This has significant implications for how we regulate activities that may harm environmental resources. A wetland system is likely to tolerate a single landowner's discarded brush trimmings. But if dumping yard debris isn't regulated, the ecosystem can be significantly degraded when all surrounding landowners decide to dump their yard waste into that same wetland. Household use of a weed killer like Roundup (which harms an array of nontarget plants and animals as well as human health[15]) on one acre of lawn quickly becomes an important issue when one hundred households on adjoining lots apply it to their properties. Streams with decreased water quality and increased flooding due to large areas of impervious surface throughout a watershed also illustrate the cumulative impact of multiple land-use activities.

Cumulative impacts may result from overuse of an area's aquifer or other water source. Pumping a well for one household may have a negligible effect on a nearby healthy stream when compared to the consequences of pumping fifty wells for as many houses in the same small subbasin. This is like a group of thirsty people crowded around one glass of water, each with a straw in the rapidly disappearing supply. Limitations on water supply—as well as drinking-water quality—may arise when too many residences, each with its individual well and septic system, are built too close together.

Preexisting environmental conditions are an important consideration when you evaluate new construction, because they illustrate what has already been lost or degraded. They set the stage for cumulative results as development activity continues. For example, the wetland acreage that remains in many places today is considerably smaller than the acreage that was originally present. According to US Fish and Wildlife Service information on the status and trends of wetland loss, by the mid-1980s, twenty-two states had lost at least 50 percent of their original wetlands; seven states had lost at least 80 percent. The majority of these were small wetlands less than five acres in size.[16] Overall wetland loss throughout a subbasin or watershed diminishes system services—for example, reducing a watershed's capacity to improve water quality or prevent flooding.

Sometimes we dismiss streams, lakes, or wetlands that have already been degraded as not really worth protection or restoration. However, a degraded wetland or stream may be more sensitive to small additional changes that can tip it over the edge of a threshold for ecosystem stability. Identifying impaired water resources allows us to plan improvements so we can avoid making a bad situation worse and presents an opportunity for restoration that may be appropriate as mitigation for new development.

Water resource impacts from the activities described in this chapter can be cumulative: small actions on small acreages add up throughout a watershed to the point where they can significantly damage watershed and ecosystem benefits and pose a hazard to human health. We will take a closer look at some examples in the next chapter.

Measuring the Impacts

> Each substitution of a tame plant or animal for a wild one, or an
> artificial waterway for a natural one, is accompanied by a read-
> justment in the circulating system of the land. We do not under-
> stand or foresee these readjustments; we are unconscious of them
> unless the end effect is bad.
>
> ALDO LEOPOLD, *A Sand County Almanac*

Four major water resource issues exemplify cumulative impacts—water pollution, flooding, water-supply depletion, and loss of species diversity. In the following pages we'll take a closer look at these impacts—and how we measure their effects.

Water Quality

Water pollution is often the first thing we think of when we consider threats to water resources. Land-use activities can change basic water chemistry in wetlands, streams, and lakes by introducing harmful chemicals, excess nutrients, organic matter (such as yard waste and woody debris), and sediment-laden runoff.

How can you tell water is contaminated? It's easier if the water sample in a jar is so murky you can't see your hand on the other side of the glass. But many of the water chemistry changes—and contaminants—in our water are invisible. Sometimes it doesn't take much; if you have ever kept an aquarium, you know that even a tiny amount of the wrong chemical accidentally added to the water can cause a mini-disaster that leaves fish belly-up on the surface. While some chemicals will kill fish outright, others are insidious, weakening fish so they are more susceptible to infection or disease. Subtle changes in water chemistry, temperature, and light can contribute to a decline in fish health, depending on species and habitat.

Many of the development activities described in the last chapter create conditions that degrade water quality—stormwater runoff; aging or inadequate wastewater-treatment infrastructure or septic systems; excess nutrients from fertilizer; toxic chemicals; grading; and removal of vegetation. Sometimes water contamination spreads throughout the interconnected waters of a sub-basin or watershed with far-reaching cumulative effects on ecosystem and human health.

Water contaminants fall into two broad categories, depending on their source. Point-source pollution originates at an identifiable location (such as a pipe carrying waste from a chemical manufacturing plant, or a boat leaking oil). Nonpoint-source pollution, carried by stormwater runoff, comprises contaminants that are washed from land surfaces into the water. Multiple sources of contamination, along with weather conditions, can combine to produce dramatic changes in lakes, wetlands, or rivers—especially those that receive water from a large watershed.

In 2011 Lake Erie experienced its most extensive algal bloom in recorded history. The stinking, green-slimed masses of algae that covered the western portion of the lake created dead zones. Decay of dead algae consumed oxygen in the water to the point where aquatic life couldn't survive. This particular type of algae (a cyanobacteria called *Mycrocystis*) produces a potent liver toxin. At the lake surface this toxin was found at levels that exceeded safe thresholds for human health by up to two hundred times. Wildlife was sickened and killed, people were sickened, recreation and all other lake use came to a dead stop.[1]

The causes of this environmental disaster can be found in the lake's watershed; they included impacts that no one had addressed, and weather conditions that made them worse:

- Sediment and fertilizers washed into the lake from urban areas and agricultural fields via stormwater and irrigation runoff. Farming practices in the watershed (earlier application of fertilizers on bare ground, less use of tillage to work it in) increased this load.
- Failing infrastructure caused the release of raw and partially treated sewage, which contributed pathogens and nutrients and decreased the water's dissolved oxygen. Combined sewer systems overflowed when overwhelmed by heavy rainfall, causing untreated sewage to wash into waterways.
- The excess nutrient load from fertilizers and sewage effluent stimulated excessive plant growth. Aquatic ecosystems require these nutrients and can cycle them, but only up to a point. Excessive plant growth changes

water chemistry by reducing the amount of oxygen dissolved in the
water. This condition is called eutrophication, and it eventually affects
all aquatic organisms.

- Atypical weather conditions changed the normal lake cycles. Unusually
heavy spring rains increased the volume of stormwater runoff and its
contaminant load. Above-average temperatures created lake conditions
that were more conducive to fast algae growth. Strong winds usually
present at that time of year were absent. These winds mix the lake waters,
bringing bottom waters to the surface and causing algae to sink to the
bottom.
- Invasive species in the lake (zebra mussels, quagga mussels) feed on
phytoplankton, reducing their numbers. Phytoplankton are minute
plants and plantlike organisms that float on top of the water and form
the base of many aquatic food chains. They normally compete with
cyanobacteria, helping to keep them under control. The invasive species
facilitated the growth of cyanobacteria by reducing phytoplankton.

Algal blooms occur in lakes throughout the United States and provide a
good example of how a range of conditions and land-use activities can cumulatively pollute our water. This glimpse of Lake Erie illustrates the importance
of evaluating watersheds as systems. We could describe each of the above
impacts individually, but we wouldn't be able to evaluate their full effect without looking at how they interact within the watershed and the lake ecosystem.

Measuring Changes in Water

To evaluate water contaminants, we need to understand some basic chemistry
and its relationship to ecosystem health. We can use the water-quality characteristics described in the biomonitoring protocol (chapter 2) to establish a
baseline for water conditions. Changes in water quality can be measured by
recording changes in these characteristics. Though developed for streams,
these basic characteristics can also be used to analyze water in wetlands, lakes,
and ponds, though the interpretation of results will vary depending on specific
wetland or lake conditions. Table 4.1 lists these characteristics, along with
examples of activities or conditions that can change them so they are no longer
within a "normal" range.

Aquatic plants and animals have different tolerances for temperature, pH,
and dissolved oxygen. Aquatic organisms thrive at optimum temperatures
according to species. Warm water holds less oxygen; sensitive aquatic species

TABLE 4.1. Selected water-quality characteristics, actions that change them, and environmental effects

Measurable water-quality characteristics	Causes of change	Examples of environmental effects
Temperature (increase)	Removal of stream-bank vegetation; high turbidity; increased impervious surfaces	Increases biochemical oxygen demand and metabolism of aquatic organisms; decreases dissolved oxygen in water; may increase aquatic organisms' sensitivity to other contaminants
pH (alkalinity/acidity)	Air and water pollution; air pollutants carried by precipitation contaminate water (sulfur dioxide, nitrogen oxides, carbon dioxide), e.g., "acid rain"	Changes physiology and health of aquatic organisms (fish, eggs, plankton, bacteria); increases toxicity of some chemicals, like ammonia, and their effect on aquatic life
Dissolved oxygen (DO)	Increased water temperature, salinity, sewage, decaying organic matter, dense algal growth; chemical contaminants; changing water depth and flows	Changes the water's ability to support plants and animals (decreased DO has negative effect)
Biochemical oxygen demand (BOD)	Disposal of yard waste; sewage; fertilizers that cause algal blooms	Depletes oxygen in water (decaying algae or other organisms use up oxygen needed by fish and aquatic invertebrates)
Specific conductance (salts)	Stormwater runoff with high salt or mineral content (road salt); airborne gases and dust transported into water via rain; irrigation return flow; water softeners	Harms fish and other aquatic organisms; contaminates drinking water; corrodes plumbing fixtures and appliances; harms agricultural crops
Turbidity (water clarity)	Stormwater runoff from eroded areas; removal of stream-bank vegetation; in-stream dredging; waste discharges; excess algae; decaying organic materials	Increases water temperature and carries disease-carrying organisms in water; interferes with photosynthesis and respiration; hinders foraging; high levels harm fish and other aquatic organisms

can survive at their "maximum" temperature only for a few hours. At certain life-cycle stages like spawning, fish have different temperature requirements. For example, the optimum temperature for spawning is seventy-seven degrees Fahrenheit for bluegill sunfish, and forty-eight degrees for brook trout. The bluegill can tolerate a short-term maximum of ninety-five degrees, while the trout can tolerate a maximum of only seventy-five.[2]

The pH of water is a measure of its acidity, which in natural systems is influenced by geology, soils, and precipitation. On a scale of 1 to 14, a pH of 1 is extremely acid, 14 extremely basic, and 7 is neutral. For comparison, vinegar has a pH of 2.2, and ammonia, 11. Most freshwater organisms prefer a pH between 6.5 and 8; tolerance and sensitivity to changes in pH vary among species. Each number represents a tenfold change in acidity; water with a pH of 5 is ten times more acidic than water with a pH of 6.

Plants and animals use dissolved oxygen (DO) present in the water for respiration; it comes from the surrounding air and from photosynthesis of aquatic plants. Warmer water holds less dissolved oxygen than cold water; altitude, season, water depth, and rate of stream flow also affect DO. Few fish can survive below a DO level of 3.0 parts per million (ppm); 6.0 is a healthy level for most fish.

Biochemical oxygen demand (BOD) measures the abundance of oxygen-consuming organisms (for example, decomposers like bacteria) in water. A high BOD results in lower dissolved oxygen and indicates organic pollution. High levels of organic material in the water increase BOD; such material includes yard waste, sewage, and fertilizers that cause algal blooms. A BOD of 1–2 ppm indicates very little organic decay and clean water; more than 10 ppm indicates unhealthy levels of decay (for example, from untreated sewage in the water).

Specific conductance is the ability of water to conduct an electric current; it measures the amount of salts and minerals that are dissolved in the water. Distilled water has a very low specific conductance; seawater has a very high level. A high salt content renders water undrinkable, kills freshwater organisms, and damages agricultural crops. But agricultural irrigation can also be a source of increased salts in water—a portion of the water used on crops evaporates or is taken up by plants, and these processes increase the specific conductance of the remaining water that reenters rivers or streams as irrigation-return flow. In streams, a high volume of flow can dilute salt content. When stream flow is low, the concentration of salt is proportionately higher, with a corresponding increase in its potential to harm aquatic life.

Turbidity is a measure of particles suspended in water, diminishing water clarity. It is measured in NTUs (nephelometric turbidity units) or JTUs (Jackson turbidity units). Water that is too turbid loses the capacity to support aquatic life. Elevated turbidity reduces the amount of light available for photosynthesis, decreasing aquatic vegetation and food for invertebrates and juvenile fish. Turbid water becomes warmer as it absorbs heat from the sun; this lowers dissolved oxygen. Turbidity interferes with the ability of fish and other aquatic predators to see their prey. Fine particulates clog the gills of fish and filter-feeding organisms, reducing growth rates, increasing susceptibility to disease, and impairing egg and larvae development. On the stream bottom, settled particulates may smother newly hatched larvae or stream insects that in turn provide food for fish.

Although sediment is a natural stream-bottom feature, and even pristine streams can run muddy during high flows, excessive sedimentation is the major cause of surface-water pollution in the United States.[3] Suspended sediment particles create turbidity and provide surfaces for attachment of contaminants such as disease organisms and heavy metals, including cadmium, mercury, lead, zinc, and chromium. These contaminants may accumulate in the food chain or concentrate on the bottom when sediment particles settle.[4]

The degree to which turbidity affects aquatic systems and organisms depends on both the amount of sediment in the water (degree of turbidity) and how long the turbidity lasts (duration of turbid flows). For example, a study reported in the North American Journal of Fisheries Management indicates that fish start to show signs of stress at 10–100 NTUs. As NTUs increase to 10,000 and exposure increases from hours to days, fish abandon cover, show increased respiration, reduced feeding rates, and increased coughing. As exposures from 10 to 100,000 NTUs lengthen into weeks, fish show reduced growth rates, delayed hatching, and long-term reduced feeding success. Long-term exposure to 500–100,000 NTUs can kill fish.[5]

Contaminants in Stormwater Runoff

Stormwater runoff, a major source of nonpoint-source water pollution, picks up contaminants from anything it flows across, including roads, junkyards, waste-disposal sites, construction sites, and agricultural fields, and carries them into streams, lakes, and wetlands. Soluble pollutants like chlorides (salts), nitrate, copper, and dissolved solids can migrate into groundwater.

Table 4.2 lists some common water contaminants in stormwater runoff, with examples of sources and effects on aquatic systems.

Some of these contaminants change water chemistry, including the characteristics described in table 4.1, which in turn affects aquatic organisms and ecosystems. Other contaminants change biological and chemical cycles—for example, the nitrogen cycle. Nitrogen is essential for plant growth. It controls the species diversity and functions of many ecosystems, which are adapted to

TABLE 4.2. Common contaminants in stormwater runoff. Source: New York State Department of Environmental Conservation, Stormwater Design Manual, 2010, table 2.1, "National median concentrations of chemical constituents in stormwater."

Stormwater runoff contaminant	Examples of sources	Examples of effects
Phosphorus	Fertilizers on agricultural crops, landscaping, lawns; detergents; animal waste; yard waste; stream-bank erosion	Depletion of dissolved oxygen; overgrowth of algae (eutrophication); harm to aquatic life and human health
Nitrogen (nitrates and nitrites)	Fertilizers on agricultural lands, landscaping, lawns; animal waste; yard waste; stream-bank erosion; fossil fuel combustion (cars, power plants); poorly maintained septic systems	Increased aquatic plant growth; algal blooms (and resulting toxic conditions); depletion of dissolved oxygen; increased decomposers (bacteria); harm to fish and other aquatic animals (high nitrite levels reduce blood's capacity to carry oxygen)
Bacteria (fecal coliform)	Animal wastes (including from pets); improperly functioning sewer or septic systems; combined sewer overflows; runoff from roads and rooftops	Harm to human health (polluted drinking water); reduced recreational opportunities; harm to aquatic life
Oil and grease	Deposition on road surfaces and parking areas (from automobiles and heavy equipment)	Reduced oxygen available to aquatic organisms; toxic effects on aquatic organisms
Polycyclic aromatic hydrocarbons	Byproducts of fossil fuel combustion	Carcinogens; harm to human health; toxic effects on aquatic organisms
Suspended solids (include silt and clay, plankton, algae, fine organic debris)	Erosion and runoff from areas where soil is disturbed or vegetation is removed; stream-bank erosion, disturbance of bottom sediment, sewage, algal blooms	Increased turbidity harmful to aquatic life; smothering of bottom-dwelling organisms; altered stream-flow patterns; conveyance of toxic chemicals (carried via suspended particles)
Dissolved solids (include calcium, chlorides, nitrate, phosphorus, iron, sulfur)	Erosion; stormwater runoff; road and pavement deicers (road salt)	Altered water chemistry; increased salinity (toxic effects on aquatic organisms; degraded drinking water)

Contaminant	Sources	Effects
Pesticides, herbicides, fungicides	Application on landscaping, lawns, agricultural areas, roadsides, medians, and rights-of-way	Impaired health and survival of nontarget plants and animals; harm to human health; ecosystem disruption
Trash and debris	Construction, industrial, and commercial sites; residential yards; dumps	Altered water chemistry and harm to aquatic life from toxic chemicals or organic matter; reduced aesthetic and recreational value of water features; impeded water flow and increased flooding
Organic matter	Runoff across impervious or eroded surfaces; dead organisms (e.g., algae); yard waste	Depleted dissolved oxygen (from decomposition); harm to aquatic life
Cadmium, copper, lead, zinc	Automobiles and heavy equipment; road salts; galvanized pipes; runoff across waste batteries and paints; corroded household plumbing systems	Accumulation in sediment; accumulation in fish and other organisms via food chain; toxic effects on aquatic life; harm to human health
Chloride (salt)	Road deicing in winter; household water softeners	Changed water chemistry; harm to freshwater aquatic species; contaminated drinking water

low levels of nitrogen, usually less than 1 milligram per liter (mg/L). Concentrations over 10 mg/L are the maximum allowed in drinking water. Excess nitrogen in water can lead to low oxygen, increased algal blooms, changes in food chains, and habitat degradation. Agricultural practices (for example, disposal of animal waste, fertilizer use), the combustion of fossil fuels, and other human activities change the nitrogen cycle; these changes contribute to the acidification of streams and lakes, decreased biodiversity, and changes in ecosystem functions.[6]

Stormwater may also carry chemicals that directly harm aquatic organisms and human health. "Hot spots" are areas that produce higher concentrations of some of these harmful chemicals, including hydrocarbons and trace metals, and contribute these contaminants to runoff:

- Auto recycling centers
- Commercial parking lots and fleet storage areas
- Landscaping nurseries and garden centers
- Orchards and crop fields (nonorganic producers)
- Heavy-equipment storage areas, industrial storage or discharge sites
- Public works areas (highway maintenance facilities)
- Dry cleaners
- Gas stations
- Petroleum storage facilities
- Wastewater treatment facilities; combined sewer outflows
- Golf courses
- Facilities that generate or store hazardous materials

Planners should site hot spots to avoid proximity to water resources, and design stormwater management so that runoff from these sites does not flow into surface water or discharge into groundwater.

How will impervious surfaces in your watershed affect water quality? Stormwater picks up contaminants as it flows across impervious surfaces and carries them into wetlands, lakes, and streams. Indicators of stream quality such as adequate flow, habitat, and water quality are related to the amount of impervious cover in the stream's subbasin. The widely recognized Impervious Cover Model developed by Schueler et al. (updated in 2009) and other related research indicates that stream health declines when impervious surfaces cover 10 percent of the land area in small watersheds (two to twenty square miles). This model finds that "stream indicators can be predicted on the basis of the percent impervious cover in its contributing watershed."[7] Based on numerous case studies, the model describes four categories of streams along a gradient from "sensitive" to "urban drainage," based on percent impervious cover:

1. Sensitive streams: Less than 5 percent impervious cover in the watershed. These streams retain good hydrologic functions and support good to excellent biodiversity. In watersheds where impervious cover is very low, indicators other than impervious cover—such as forest cover, road density, agricultural practices, and buffers—should be used to evaluate stream quality.

   Transition from sensitive to impacted: 5–10 percent impervious cover. In general, a high-quality stream has less than 10 percent impervious cover in its watershed and can continue to function and support good to excellent diversity of aquatic life.

2. Impacted: 10–20 percent impervious cover. These streams show signs of declining health. Most stream health indicators are in the "fair" category. Streams with extensive riparian vegetation tend to be healthier.

   Transition from impacted to nonsupporting: 20–25 percent impervious cover.

3. Nonsupporting: 25–60 percent impervious cover. These streams do not retain their original hydrologic functions, channel stability, habitat, water quality, or biodiversity. They are degraded to the point where it may no longer be possible to restore them to their predevelopment condition.

   Transition from nonsupporting to urban drainage: 60–70 percent impervious cover. Streams can no longer support stream habitat, water quality, and biodiversity. These streams are so degraded that they can't be restored to predevelopment conditions.

4. Urban drainage: More than 70 percent impervious cover. These highly modified streams have very poor water quality and function only as water conveyance channels.

These general categories were designed for use with actual stream monitoring data. Since they were developed for relatively small streams, they may not apply to larger streams with major point-source pollution or impoundments and dams.

Relationships between impervious cover and water quality can guide the evaluation of land-use activities. Most studies find that water quality begins to decline even at very low impervious cover percentages, depending on the overall watershed condition. General consensus among researchers is that at

10 percent impervious cover, stream health is impaired; by 25 percent, streams are degraded and can no longer support aquatic organisms (such as certain fish, insects, amphibians) that were originally present. Some sensitive species like brook trout may have trouble surviving in watersheds where impervious cover is below 10 percent (studies document the absence of brook trout in some watersheds with as little as 4 percent impervious cover).[8] EPA and Center for Watershed Protection websites list studies that document relationships between percent impervious cover and a variety of physical, chemical, and biological characteristics, including stream channel stability, depth, base flow, presence of sensitive fish, and aquatic invertebrates. This relationship between impervious surfaces and water quality applies to wetlands as well; one study noted changes in wetland water quality when impervious surfaces within the contributing drainage area reached 3–5 percent; impervious surface cover of more than 20 percent resulted in degraded wetland water quality.[9]

When you compare these results with the impervious cover associated with land-use categories (listed in table 2.1), it is evident that land development can exceed 10 percent or more impervious cover fairly easily—this includes most residential zoning for lots that are two acres or smaller. An entire subbasin zoned for land use that exceeds 10 percent impervious cover is likely to change the quality of the stream that drains it. This relationship underscores the importance of using stormwater management methods based on "best practices" to keep impervious cover percentages in a subbasin as low as possible and to treat runoff before it enters water resources.

Stormwater carries contaminants that can soak into the soil, entering the water table and groundwater. Where the water table is at or near the ground surface, as in many wetlands, or intersects with a stream or lake bottom, contaminants may enter groundwater directly from land or surface waters. Because groundwater–surface water interconnections may be difficult to see or measure, you may not notice water-quality changes immediately. Once contaminated, groundwater cannot be cleaned up; if it is lost as a source of drinking water, it cannot be restored.

Toxic Chemicals

We employ an array of chemicals in all aspects of our daily lives, but when they escape into our water, their effects are anything but beneficial. These chemicals can cause changes in ecological systems, harm aquatic organisms, and pose hazards to human health. Our waters receive an array of chemical contaminants from stormwater runoff, commercial and industrial waste discharges (including those from energy development), household waste (cleaning products and pharmaceuticals), agricultural operations, and spills and leaks associated with pipelines and the transportation or storage of waste materials. In 2009, industrial facilities alone released more than 200 million pounds of toxic chemicals into US waters.[10]

In 1976, Congress passed the Toxic Substances Control Act (TOSCA) to allow the EPA to regulate existing chemicals that may harm human health or the environment, and also new commercial chemicals before they enter the market. Despite the large number of chemicals available (fifty thousand to eight-five thousand, depending on source), TOSCA doesn't require chemicals to be proven "safe" before allowing their use. The EPA is required to establish that a suspect chemical is an "unreasonable risk of injury to human health or the environment" before it can require extensive toxicity testing for that chemical. As a result, few chemicals have been tested and restricted. By 2010, the EPA had required testing of about three hundred chemicals and restricted the use of only five.[11]

Chemical companies manufacture more than one million pounds of approximately twenty-five hundred "high production volume" (HPV) chemicals annually; nearly 45 percent have not been adequately tested to evaluate their health effects on humans and wildlife.[12] EPA data similarly indicate that most chemicals haven't been tested.[13] Many of the chemicals that have been tested do harm human and ecosystem health—benzene, arsenic, formaldehyde, lead, mercury, glyphosate, and vinyl chloride are examples. Some of these cause cancer (carcinogens); others affect the nervous system. Endocrine disruptors from a variety of agricultural, industrial, and domestic sources (for example, pharmaceuticals) "disrupt internal biological processes such as development, growth, and reproduction that are regulated by hormones. Whether these compounds are present at sufficient levels in our waterways to harm human health remains a topic of serious concern and ongoing research."[14]

The EPA does not require drinking-water treatment plants to test for many chemicals; some of these are toxic, and others are of unknown toxicity. Treatment plants are not designed to remove many industrial chemicals, road salt, and naturally occurring radioactive materials (NORM) that pass through and are discharged into rivers and streams.

In recent years scientists and citizens have expressed concern about the presence of chemicals from personal care (cosmetic) and pharmaceutical products (prescription and over-the-counter drugs, veterinary products) in our streams and rivers. "In 2008, the Associated Press found an array of pharmaceuticals, from pain killers to antibiotics to mood stabilizers, in the drinking water of 24 major metropolitan water suppliers. Even worse, 34 of the 62 water suppliers contacted by the AP couldn't provide results as they had never tested for pharmaceutical compounds."[15]

Contaminant Levels and Testing

At what point is a chemical harmful? Concentrations at which chemicals affect people or aquatic organisms are often difficult to determine. Regulatory thresholds for chemical safety are based on doses that will kill an adult; we don't know what dose will harm a child, much less kill a fish. Some of these chemicals, like endocrine disruptors, produce subtle effects that make you sick over time by disrupting the way your body system works. Similarly, chemicals may interfere with water chemistry, biological functions, and eco-systems.

The EPA website provides a list of drinking-water contaminants, their potential health effects, and maximum allowable levels. Contaminants fall into specific groups: microorganisms, disinfectants and their byproducts, inorganic chemicals, organic chemicals, and radionuclides.[16] In looking at these lists it is important to read all definitions and footnotes carefully. The regulations describe two types of maximum contaminant levels:

1. Maximum contaminant-level goal (MCLG) is the level of contaminant in drinking water below which there is no "known or expected" risk to health. This standard allows for a margin of safety and represents a public health goal—which is not enforceable.
2. Maximum contaminant level (MCL) is the highest level of a con-taminant that is allowed in drinking water. While these standards are enforceable and are set "as close to MCLGs as feasible using the best available treatment technology," they also take economics into account—that is, the cost of complying with, and enforcing, the regulations.[17]

The significant distinction between these levels is that although the con-centration of a chemical may be within legal limits, it may also be well above (nonenforceable) public health guidelines. Regulations established to protect drinking water from industrial and agricultural pollutants are subject to questions about the level of protection these standards actually provide. "For many chemicals, two numbers exist: the enforceable maximum and the health-based maximum-contaminant level goal. . . . The enforceable values for the carcinogens benzene, vinyl chloride, and trichloroethylene, for instance, have been set at 5, 2, and 5 parts per billion, respectively. Their maximum-contaminant-level goals, however, are all zero."[18]

The EPA provides information about two other categories of drinking-water contaminants:

- *Secondary Drinking-Water Regulations.* These are nonenforceable guidelines for regulating chemicals that may cause cosmetic changes (for example, skin discoloration) or aesthetic effects (water taste, odor, color). States have the option to adopt these as enforceable standards.
- *Unregulated Contaminants.* This is a list of contaminants that are not subject to national primary drinking-water regulations but that are known or suspected to occur in drinking water. For more information see the list, or visit the Drinking Water Contaminant Candidate List (CCL) website.

When we consider the effects of water contaminants on children who may drink the water they are swimming in, the line between "drinkable" water and recreational water blurs and calls into question the difference in standards for water quality depending on best use. It's a good idea to keep up to date on the latest regulations and thresholds for water quality in your area. In the spring of 2013, the EPA decreased the safe level threshold for fecal coliforms. In the Hudson River, many swimming areas may meet the new standard but nevertheless pose a health risk, especially after storms (largely because of overflow of combined sewer systems along the river). Waters in your area may need additional testing before you can thoroughly evaluate their condition.

For additional information about harmful levels of contaminants in water that may pose a significant threat to human health, ecosystems, or livestock, refer to websites—including state health department or water-quality protection agencies, research institutions, and medical reports. The Endocrine Disruption Exchange, a nonprofit organization, is "dedicated to compiling and disseminating the scientific evidence on the health and environmental problems caused by low-dose exposure to chemicals that interfere with development and function."[19] The Environmental Working Group (EWG)[20] describes chemical contaminants in drinking water that are currently not regulated even though they pose a threat to human health. In 2009, EWG data revealed that a number of US cities have drinking water with unhealthy levels of contaminants, in some cases exceeding recommended health thresholds. Test results from EWG's database covered 316 contaminants in water supplies of communities in forty-five states, including 202 chemicals that are not subject to any government regulation.[21]

Testing services can give you valuable information about your water. Water-testing apparatus—for example, Hach or LaMotte water test kits and measuring devices—are available from a variety of sources; check USGS, EPA, or state water-quality monitoring websites for additional recommendations. States or counties test most public wells for the presence of bacteria, an immediate health risk. But other contaminants like uranium and arsenic are not apparent in water and are not routinely tested. Companies like National Testing Laboratories LTD in Ohio[22] provide customized water tests for more than one hundred contaminants, including bacteria, heavy metals, inorganic contaminants, organic chemicals (many of those found in pesticides), as well as physical water characteristics like pH and turbidity. Water-quality testing is more accurate when samples are taken from the entire water column, as well as sediment, because the characteristics of various contaminants may cause them to sink to the bottom, or circulate at different water depths.

Property owners are generally responsible for private water-supply testing. You can have water tested for particular contaminants; if you know the type or source of pollution or suspect your water is unsafe, you may obtain help from a local health department or state agency (cooperative extension, public health, water resource managers) or professional water-quality consultants. In New York State, for example, the Department of Health generates a list of certified laboratories. The Cornell University Cooperative Extension provides detailed information on water-quality testing and contaminants (water quality information for consumers at http://waterquality.cce.cornell.edu). The EPA supplies water-quality information for other states; refer to the fact sheets and brochures on the EPA website (http://water.epa.gov/drink/contaminants/index.cfm).

Flooding and Watersheds

Floodwaters are on the rise. You have probably noticed more flooding where you live, even along small streams and in places where flooding wasn't an issue years ago. Floods are now the number-one disaster in the United States.[23] People who live outside mapped high-risk flood areas file nearly one-fourth of all National Flood Insurance Program (NFIP) claims. In 2012, total claims paid by NFIP totaled $7.7 billion.[24] Floods are caused by a variety of conditions, including more frequent, severe storms due to climate change; increased impervious surfaces; construction within floodplains; wetland loss; and reduced forested cover and buffer vegetation. When we clear vegetation, grade a site level, and fill in small wetlands and streams, we change natural drainage patterns and disrupt the watershed's built-in flood-control system. We then have to build replacements for lost watershed services. When we fill a wetland or small stream, we need to construct a stormwater basin or install culverts and pipes. If we build within a floodplain, we need to flood-proof buildings and divert water flows to protect lives and property. We can reduce some of the costs of these actions if we retain natural watershed services whenever possible.

Impervious Surfaces

According to the US Department of Agriculture, between 1945 and 1997, urban land in the United States increased by about 327 percent; the EPA reports that paved-road mileage during that time increased by 278 percent.[25] As our activities cover more land with impervious surfaces, the volume of runoff increases. The US Government Accountability Office provides a general example: 90 percent of the precipitation that falls on land with natural ground cover infiltrates the ground, and only 10 percent becomes runoff. However, when impervious surfaces cover 75 percent of the ground, 55 percent of the precipitation is runoff, and on paved parking lots this rises to 98 percent.[26] In a forested watershed, 10 percent of precipitation is runoff, 40 percent is lost through evapotranspiration, and 50 percent infiltrates the soil. When impervious surfaces in the watershed increase to 10–20 percent, evapotranspiration decreases slightly, runoff increases to 20 percent, and infiltration decreases to 42 percent. When impervious cover is more than 75 percent, 55 percent of precipitation is runoff, 30 percent is lost to evapotranspiration, and only 15 percent infiltrates the soil.[27] These percentages will vary depending on types of vegetation, soil, and local conditions, but the main point is that impervious surfaces dramatically change the volume of stormwater runoff.

Impervious cover is associated with all the impacts discussed in this chapter: water quality, increased runoff and flooding, decreased groundwater recharge, and decreased biological diversity.[28] Increasing the impervious cover in a watershed triggers a series of effects. Stormwater runoff increases (for example, a one-acre parking lot can generate sixteen times more annual stormwater runoff than a one-acre meadow).[29] This runoff carries a substantial load of both sediment particles and contaminants. Stormwater runoff changes the physical, chemical, and biological characteristics of healthy ecosystems—which in turn can lead to increased downstream flooding, erosion, and enlargement of the stream channel. As stormwater flows across impervious surfaces, less water soaks into the ground for groundwater recharge, and water tables may drop—lowering water level in wells and reducing stream flows, especially during drought.

Climate change and weather patterns, as well as land-use activities that change where water is stored and how it moves through a watershed, influence these negative effects.

In summary, impervious surfaces can amplify flooding by

- increasing the amount of stormwater runoff;
- changing stormwater flow destination;
- increasing peak flow from storms (flooding downstream areas);
- decreasing time to peak flow (creating "flash" streams);
- increasing length of time for peak flow;
- changing wetland hydroperiod, water levels, and flood storage capacity; and
- covering or filling in small wetlands and streams.

Network of Wetlands and Small Streams

Wetlands and small streams throughout a watershed catch runoff and precipitation and store it or move it downstream. The EPA estimates that the United States had over 220 million acres of wetlands in the 1600s. Today less than half remain.[30] In general terms, if a one-acre wetland, one foot deep, can hold approximately 330,000 gallons of water, and twenty-five of these wetlands are scattered throughout a watershed, that is a potential water storage capacity of 8.25 million gallons. If we fill wetlands or decrease their storage capacity, we diminish the watershed's ability to absorb excess stormwater flows and floods.

According to a US General Accountability Office study, watersheds with less than 10 percent area in wetlands have higher peak stormwater flow. Some watersheds retained flood control and general water-quality functions when at least 3–7 percent of their area was wetland; they needed 15 percent wetland cover to effectively remove phosphorus from runoff.[31] Protecting and restoring wetlands can reduce the destructive potential of flooding and thus serve as an important component of a comprehensive flood-protection strategy. Activities that disturb the physical characteristics of wetlands (clearing and grading, filling, draining) change their capacity to store water by changing hydroperiod, water source, vegetated buffer, and land-use condition within the wetland's contributing drainage area.

A system of small streams throughout a watershed intercepts precipitation and runoff and conveys water through a network of channels, like capillaries that convey blood to larger blood vessels. Our cardiovascular system depends on billions of capillaries for healthy functioning;[32] similarly, a network of small streams is vital to watershed health. EPA estimates that 55 percent of the streams that supply drinking water to NY state residents are small intermittent, ephemeral, or headwater streams. This system is better able to dissipate flooding from heavy rainfall, as more water "conveyances" are available to catch and transport water. When these small streams are filled in or diverted, the watershed loses its capacity to catch water, and a larger volume of runoff washes directly into larger streams and rivers.

Watershed Vegetation

Watershed health is influenced by vegetation, especially the amount of forested cover along streams. Plant cover holds soil in place, slows flood flows, takes up water through root systems, and releases water back into the atmosphere. It reduces runoff, allowing water to seep into the ground where it falls. The type of plants and their location within a watershed and in relation to streams, wetlands, and lakes influence how well these systems reduce flooding.

Vegetated buffers along the edges of water resources contribute to natural flood control. More than 80 percent of riparian habitat in North America has disappeared over the past two hundred years; adjacent or upstream human activities constantly threaten remaining areas. Riparian areas account for a small proportion of the landscape but are "regional hot-spots that support a disproportionately high number of wildlife species and provide a wide array of ecological functions and values."[33] The value of riparian areas is especially evident in regions with less rainfall where forests typically aren't supported, or where much of the original forested cover in a watershed has already been lost.

A healthy stream has forested cover along its edges, on both sides. Removal of a significant percentage of watershed forests and trees along the edges of surface waters leads to

- increased bank and stream channel erosion;
- increased volume of stormwater runoff;
- loss of natural contaminant-removal services; and
- increased downstream flood flows.[34]

The headwaters of streams and rivers, which are often springs, seeps, or wetlands, are especially sensitive to loss of forest (or other native vegetation) because the water that flows from these sources (and the nutrients it carries) can affect all areas downstream. Research that documents the importance of forested watersheds, buffers, and riparian areas is available online.[35] The USDA publication *Conservation Buffers: Design Guidelines for Buffers, Corridors, and Greenways* is a useful guide.[36]

Floodplains

When connections between streams and their floodplains are not obstructed, floodwaters that top the banks spread overland and slow down. Even very small streams with floodplains are important to overall watershed flood reduction. When you build within floodplains, grade them, or remove native vegetation, you decrease their ability to provide protection during floods. Floodplain construction increases the risk of downstream property damage and the threat to general safety; it increases runoff, contributing to the flow and force of floodwater. When we lose the natural flood-protection services of intact floodplains, we have to engineer replacement services, whether through reinforcing structures to make them less prone to flood damage, reinforcing banks, building berms and other barriers to floodwaters, or redirecting floodwaters to another location.

Development and land-use activities within floodplains can aggravate existing flood conditions, put more people in harm's way, or break down natural watershed-protection systems. In summary, these activities include

- reducing or eliminating buffers;
- filling riparian wetlands and small streams;
- removing native floodplain vegetation;
- separating the stream channel from its floodplain;
- degrading banks;
- dredging or straightening stream channels;
- increasing erosion and sediment deposition; and
- increasing impervious surfaces within floodplains.

A watershed-wide approach should address potential impacts to floodplains and include multiple actions that protect the watershed system, its streams, buffers, and wetlands, so that it can perform optimally over the long term.

Groundwater Pumping and Changes in Water Supply

The long-term sustainability of groundwater and aquifers is influenced by climate change and weather extremes (hot to cold, deluge to drought); decreased groundwater recharge; and increased water use for residential, commercial, agricultural, and industrial purposes. Water use and drought not only deplete drinking-water supplies but also decrease the water necessary for sustaining aquatic ecosystems. We reduce the amount of precipitation available to replenish groundwater when we increase impervious cover within a watershed, direct water from one subbasin to another, remove vegetation, and compact the soil.

Water withdrawal changes the local groundwater-flow system. The cumulative pumping from many wells may have important regional effects not only on groundwater supply but also on water levels in streams, rivers, and wetlands—especially in regions of low rainfall. When water is pumped from a well, groundwater levels at and near the well decrease. Water moving slowly underground toward the place where it discharges into a wetland or stream is diverted to the well area. When we pump water from a groundwater supply, we have to replace it to maintain stable groundwater storage levels; if we do not, underground water reserves are used up over time.

Development projects often produce a "water budget," which describes natural conditions before construction begins. Also called "safe yield," this budget calculates the amount of water available for human consumption. This means that groundwater pumping for human use is considered "safe" if this water withdrawal doesn't exceed the natural rate of groundwater recharge. USGS considers this safe-yield method problematic: "This concept has been referred to as the Water-Budget Myth . . . because it is an oversimplification of the information that is needed to understand the effects of developing a ground-water system."[37] To maintain the groundwater storage level over time, we must compensate for water removed from the groundwater storage system either by increasing groundwater recharge or decreasing groundwater discharge, or by a combination of both. In addition, the effects of drought and climate change must be factored into any assessment of the amount of groundwater available for human use. According to the USGS, "How much groundwater is available for use depends on how these changes in inflow and outflow affect the surrounding environment and what the public defines as undesirable effects on the environment."[38]

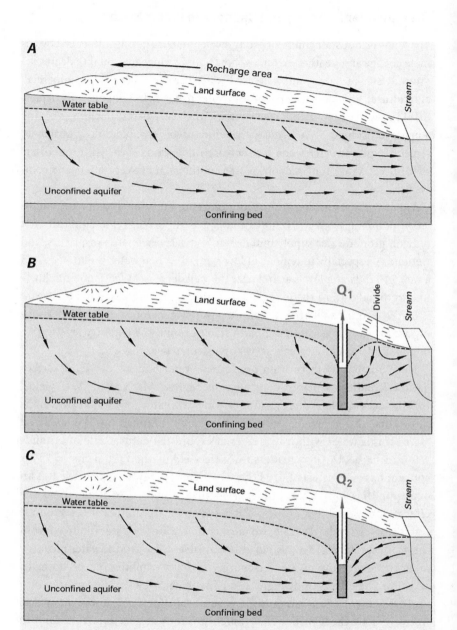

FIGURE 4.1. Effects of groundwater pumping. From Thomas Winter et al., *Ground Water and Surface Water: A Single Resource*, US Geological Survey Circular 1139, 1998. Courtesy of Department of the Interior, USGS.

Figure 4.1 shows three scenarios that illustrate effects of pumping on groundwater and a stream:

A. Under natural undisturbed conditions, groundwater discharges to a stream.
B. A well pump intercepts some of the groundwater that would have gone to the stream (Q1 is the rate of pumping) and changes its flow direction.
C. If the rate of pumping increases (Q2), it not only intercepts groundwater before it reaches the stream but also draws water from the stream to the well.

Increased water use and more wells reduce the amount of water available for streams, lakes, and wetlands by changing flow, depth, and hydroperiod. Groundwater pumping can reduce stream flow below the level that fish and wildlife require, especially during times of drought, and lead to changes in water quality. Longer periods of low flow can damage vegetation along the stream, change bank stability, increase water temperature, reduce dissolved oxygen levels, and alter patterns of sediment deposition. Subsequently, these changes can harm aquatic plants, insects, and fish. Reduced stream or river flow may also impair recreational use, reduce the ability of a river to assimilate wastewater discharges and other contamination, concentrate pollutants and their effects as water level drops, and reduce water available for agricultural irrigation.

In wetlands, hydroperiod determines which plants and animals thrive and how much water is stored or conveyed seasonally to streams and other interconnected waters. Groundwater pumping may lower groundwater levels near wetlands, changing wetland hydrology, vegetation, and habitat value. These changes are especially significant during the growing season, leading to increased invasive species and decreased biodiversity, and impairing the wetland's ability to process contaminants.

To understand the local effects of groundwater pumping, consult local precipitation records, well data, and agricultural and residential water-use records. You may require the services of an experienced hydrologist to evaluate local water use and its effects on groundwater and surface waters. To identify local impacts on water supply, consider the following:

- Average human water-consumption levels
- Current water demand and projected demand over time
- Climate change or drought conditions
- Cumulative effects of multiple wells on stream flow and hydrology
- Minimum stream flow required to sustain an aquatic ecosystem
- Ratio between projected groundwater use and the aquifer recharge rate

Groundwater pumping changes water conditions above and below the ground. To develop thresholds at which the level of (water supply) change becomes unacceptable, examine the tradeoffs between groundwater use and its consequences for surface waters and ecosystems. The USGS supplies detailed groundwater information and easy-to-follow illustrations.[39]

Groundwater depletion is an important indicator of regional water stress. At the 2013 World Economic Forum in Davos, Switzerland, experts identified "water risk" (which includes floods and drought) as one of the top four risks facing businesses this century.[40] Responding to the need for data, the World Resources Institute compiled the Aqueduct Water Risk Atlas, which uses twelve indicators to map areas where water risk may be a problem.[41] Overall risk is presented as a number, from low to extremely high. Risk factors for water stress include the following:

- Competition for water (water withdrawals as a percent of total available water)
- Variation in water supply (annual, seasonal)
- Number of past floods (from 1985 to 2011)
- Drought: length of time, severity, frequency
- Upstream water-storage capacity and the ability of upstream areas to buffer flood and drought
- Groundwater stress: ratio of water withdrawal to rate of recharge
- Protected lands: percent of water supply originating in watershed area protected from actions that degrade water quality
- Threatened amphibians: presence of threatened species, indicating sensitive ecosystems

In the United States, areas with the highest levels of water stress are along the East Coast, especially metropolitan areas from Washington, DC, to Boston; along the Great Lakes, particularly Lake Michigan; and portions of western states (Texas, Oklahoma, Kansas, and states to the west).

This atlas project gives us specific data about water availability as a guide for assessing water-use impacts.

Decreased Biological Diversity

Changes in water quality, availability, and depth alter plant and animal habitats. If these changes are significant, some species won't be able to survive. Changes in the variety of species present in a habitat influence primary productivity, nutrient cycles, predator-prey relationships, and other ecosystem services. We may not immediately notice these impacts, especially if they take place gradually, involve species that we don't usually see, and are cumulative.

Asking the "so what" questions explained in chapter 3 may help you identify a progression of environmental changes and their consequences. Sometimes the consequences are sudden, as when a toxic chemical spills into the water and kills fish. But often, ecosystem changes occur along a gradient from pristine to degraded conditions—for example, as a forested swamp habitat is drained, then cleared, and finally planted in agricultural crops; or as a few stalks of Japanese knotweed sprout on a riverbank and eventually crowd out all other vegetation.

Natives and Invasives

Introduction of nonnative invasive species to an ecosystem can produce dramatic changes due to a corresponding loss of native plants and animals and their roles in keeping the ecosystem healthy. Often we don't immediately know the long-term result of replacing one species with another, or the cumulative consequences of losing multiple species in an ecosystem. We can picture the ecosystem as a Jenga tower—the game where the object is to remove one block at a time from the structure without toppling it. Removal of some blocks weakens the position of other blocks, and sometimes it's hard to predict which removal will weaken the structure to the point where it literally falls apart. If we can't describe the consequences of removing this or that block from the tower, or species from an ecosystem, the best way to prevent system "toppling" is to refrain from removing anything—or at least to proceed with great caution.

Nonnative species tend to invade when an ecosystem is degraded, because they are better able to tolerate habitat disturbances or changes. This replacement often reduces the total number of species present (biodiversity).

For example, Japanese knotweed, an ornamental from Asia, has the capacity to overwhelm riparian or wetland areas and native plants; it can survive severe floods and quickly colonize scoured or disturbed areas. It grows from plant pieces dumped into rivers as yard waste, roadside mowing, construction projects, or fill soil from riparian areas. Flooding spreads it rapidly throughout a floodplain or riparian area. Its dense thickets decrease habitat value for wildlife, alter soil chemistry and nutrient cycling, and shade out other plants, including young trees that try to sprout beneath it. Knotweed thickets reduce abundance and diversity of insects, decrease wildlife habitat quality, and increase the risk of stream-bank erosion as the invasive replaces woody species.

Purple loosestrife invades wetlands, changing nutrient cycling, reducing plant diversity, and degrading wetland habitat suitability for specialized wetland birds (such as black terns, least bitterns, pied-billed grebes, and marsh wrens).[42] Dense stands of purple loosestrife crowd out native plants, changing physical habitat (for example, reducing shallow water areas needed for fish spawning and foraging) and food sources (waterfowl do not eat it). Biological control by introducing insects that eat purple loosestrife may eventually restore native species diversity to treated areas and reduce the use of harmful chemical herbicides.

Native insects are the often-overlooked "foundation" of many ecosystems; a variety of insects is critical for healthy natural systems. Insects are an important source of food for many other animals, including other insects, and are an essential source of protein for nestling birds. When we clear and grade an area for development and remove the native plants, a source of food for insects is gone. Replacing native plants with nonnative species commonly used in landscaping does not restore this food supply, with dramatic consequences for breeding birds and other animals that depend on native insects for food. Most plant-eating insects are specialists: they eat the leaves of only a few types of plants. The most specialized eat foliage from only one plant species; if the plant is removed from an ecosystem, the insect will no longer thrive.[43] No two species of plants have the same leaf chemistry, so plants vary in their digestibility and toxicity to insects. Native insects aren't adapted to nonnative plant chemicals (that is, they can't digest the leaves); as a result, because they have few predators, these plants may outcompete native species.

Doug Tallamy's Bringing Nature Home illustrates the effect of nonnative plants on butterfly and moth caterpillars: "In a comparison of . . . [butterfly and moth] larvae produced on native and alien woody plants in Oxford, Pennsylvania, native plants supported 35 times more caterpillar biomass, the preferred source of protein for most bird nestlings, than alien plants supported."[44] Native oak trees supported the highest number of caterpillar species (as many as 534), closely followed by willow, cherry, and birch. Nonnative ornamentals lagged far behind in the number of insect species supported.[45]

Trees are not interchangeable in terms of their ecosystem value: this applies to their status as native or nonnative and also to their growth form. Planting ten Bradford pears will not replace the habitat value of ten oaks. Similarly, ten oaks less than five years old will not replace the value of ten oaks that are more than fifty years old. Removing mature forest and replanting with saplings does not replace lost habitat value.

Though native insects illustrate the complexities of ecosystems, they are only one group of living things to consider when we look at changes in biodiversity. Development activities, chemical contaminants, and changes in water depth can change the number of plants, microorganisms, birds, fish, amphibians, and other groups that inhabit an ecosystem. These changes can lead to loss of bank or shoreline stability, altered nutrient cycling, decline in productivity, and loss of other system services.

Species Changes as an Indicator of Ecosystem Health

Threats to plant and animal diversity in an ecosystem can be difficult to measure. Several approaches use an inventory of species to evaluate ecosystem condition. For example, the rapid bioassessment protocol (described in chapter 2) uses an inventory of macroinvertebrates in a stream as an index of stream condition along a gradient from healthy to degraded. The organisms used for this determination are divided into groups depending on how well they can tolerate pollution. Water-quality monitoring measures the relative numbers of individuals of species that are pollution-tolerant (including blackflies, midges, and aquatic worms) and those that are pollution-sensitive (including mayflies, caddisflies, and stoneflies). Pollution-sensitive organisms tend to require high dissolved oxygen levels. When present in large numbers, these macroinvertebrates indicate the stream is in good condition.

The focal-species approach developed by the Wildlife Conservation Society's Metropolitan Conservation Alliance (MCA) project uses field survey data to compare wildlife species that respond differently to development impacts.[46] MCA's focal-species method uses two wildlife groups: birds, and reptiles/amphibians because they respond to development pressure and habitat fragmentation and are good indicators of ecosystem condition. Amphibians are sensitive to water contamination absorbed through the skin and are reliable indicators of aquatic habitat quality.

The method divides species into two broad categories:

1. *Development-sensitive species.* These are habitat specialists that require specific habitat conditions or a combination of different habitat types. When these habitats are lost or degraded, populations of these species decline.
2. *Development-associated species.* These are habitat generalists, adaptable to a range of conditions and with less specific habitat needs. They can tolerate habitat disturbance or degraded conditions, and their populations can flourish in developed areas.

The model compares the proportion of development-sensitive to development-associated species to determine the relative quality of a habitat or ecosystem. As land is developed, species that can tolerate development replace development-sensitive species. The result is fewer species overall and a loss of biodiversity. A shift in species parallels the gradual change from pristine to degraded habitat.

TABLE 4.3. Comparison of development-associated with development-sensitive species in a northern Wallkill Valley, NY, study

	Amphibians	Reptiles	Birds
Total number of species surveyed	17	15	85
Development-associated species	6	5	13
Development-sensitive species	11	10	72
NYS listed species* that are also development-associated	0	0	0
NYS listed species* that are also development-sensitive	3	5	9

Data from Danielle LaBruna and Michael W. Klemens, *Northern Wallkill Biodiversity Plan: Balancing Development and Environmental Stewardship in the Hudson River Valley Watershed*, MCA Technical Paper 13 (New York: Wildlife Conservation Society, 2007).

Species listed by New York State as endangered, threatened, or special concern.

In the northern Wallkill Valley region of New York's Hudson Valley, a focal-species analysis found that "development-sensitive species, such as the spotted salamander and wood frog, which are now rare in more southerly New York locations, are still abundant in . . . [this] region. However, at the same time that highly development-sensitive species were observed, many development-associated species . . . were observed as well. This indicates that this landscape, although still rural, is showing the signs of degradation from sprawl's influence."[47] Table 4.3 summarizes findings from this study.

While the species information in table 4.3 is a highly simplified excerpt from a larger study, it illustrates a general correlation between species diversity and land development. The more extensive the development, the fewer species present. Approximately 80 percent of the 117 species surveyed in this study are sensitive to development. The "development sensitive" category includes all the species listed as threatened, endangered, or of special concern.

Biodiversity in a region declines one species at a time, one habitat at a time. Small wetlands and streams are especially vulnerable to the consequences of species loss. This incremental loss in small areas adds up until it eventually influences larger ecosystems and watersheds. The cumulative result is ecosystem change that pushes the "condition" gradient from pristine toward degraded. Degraded ecosystems with reduced species diversity don't produce the services and benefits of healthy ecosystems. Development-induced changes in biodiversity begin at the local level, and you can apply solutions at that level as well. Protecting ecosystems and paying attention to local biodiversity begins in your community, its habitat corridors and backyards. Local protection of the natural systems that support a diversity of species can have positive cumulative effects on regional watershed health.

Energy Development
and Water

In short, a land ethic changes the role of *Homo sapiens* from con-
queror of the land community to plain member and citizen of it.
It implies respect for his fellow members and also respect for the
community as such.
ALDO LEOPOLD, *A Sand County Almanac*

We depend on a reliable supply of energy. Yet the type of energy we use,
and how we obtain it, have become sources of controversy and mag-
nets for misinformation. Everyone wants energy development to be clean and
safe, but wanting doesn't make it so. When a pipeline leaks or toxic wastes are
dumped into a stream and kill the fish, the event gets media attention. But it
may be hard to uncover all the facts.

In chapter 3 we looked at some of the impacts associated with residential
development and agricultural activities. Energy development projects share
some of these activities and produce some of the same effects, on a larger
scale. They also produce a new set of impacts on watersheds and ecosystems.

While many types of energy development can affect water resources, this
chapter focuses on natural gas, oil, and coal (fossil fuels) and is designed to help
you understand the energy development projects you read about in the news.
Fossil fuels have earned a reputation for environmental contamination. Often
media reports about hydrofracking, oil drilling, pipelines, or coal mining don't
give you enough information to understand environmental (water resources)
impacts. The following pages offer an approach to evaluating complex energy
development projects by breaking them down into manageable parts so you can
evaluate their effects more accurately and easily. Then you can consolidate this
information to examine cumulative impacts and true costs of development.

Controversy about energy sources involves issues that can divide communities:

- Concern for personal health and environmental protection
- An urgent need for jobs and local economic health

- The need for reliable, affordable energy to meet a steadily growing demand
- The challenge of connecting energy development activities with effects on the environment (as documented by scientific facts)
- The power of corporate interests

We need to approach energy development with an eye toward ferreting out the facts, weighing the true costs, and adopting solutions that don't force us to choose between energy and water.

Conflicting "Facts"

When we read the local newspaper, listen to radio, watch TV, or access the Internet, we are bombarded by snippets of information and opinions that often contradict one another. The following news excerpts about the high-profile proposed construction of the Keystone XL pipeline, which would transport oil from Canadian tar sands in Alberta to refineries on America's Gulf Coast, illustrate this "snippet" reporting. The ultimate fate of the pipeline is not the point here; these news clips only serve to illustrate the political nature of the issue and the clash of conflicting information. The first is from CNN:

> Republicans called on President Barack Obama to make a decision on the controversial Keystone XL pipeline in their weekly address Saturday, saying the delay was a result of political calculations. Speaking for his party, Wyoming Sen. John Barrasso said the project has received wide support, except from Obama's political base.
>
> "Everyone from members of the United States Chamber of Commerce to members of labor unions support this project," Barrasso said. "But the President has threatened to veto this bill because the pipeline is opposed by a number of extreme environmental groups. These are the same groups who in the past have supported the President and he needs their support for his reelection."
>
> Barrasso asserted it was "time for the President to stop playing politics."[1]

The second is an op-ed from the *New York Times*:

> It looks increasingly likely that the State Department will approve the Keystone XL pipeline. . . . An existing pipeline carrying tar sands oil . . . was forced to shut down for repairs after springing two leaks last May in North Dakota and Nebraska. That is one reason why Dave Heineman, Nebraska's Republican governor, has asked that the new pipeline be rerouted. He fears a

spill could pollute the Ogallala Aquifer, a crucial water source beneath the Great Plains.

. . . The State Department appears to be more persuaded by proponents who claim that the pipeline will help reduce America's dependence on oil from politically troubled sources in the Middle East. . . . What pipeline advocates—including big-oil lobbyists and House Republicans who have tried to force an early, favorable decision—fail to mention is that much of the tar sands oil that would be refined on the Gulf Coast is destined for export. . . . TransCanada has said the 2,000-mile line would create 20,000 jobs in the United States. The State Department concludes that the real number may be closer to 6,000.[2]

What do you learn about this project based on these two articles? They reveal that the issue has become political for some elected officials—whether you are for or against it depends on which political party you support. Yet although Republicans in Congress support the pipeline, a Republican governor opposes a proposed pipeline route. "Extreme" environmental groups oppose pipeline construction. Possible pollution of an aquifer is an issue. Pipelines leak. Job creation is an issue, but estimates of how many jobs vary considerably, depending on who is providing the figures. Those who support the pipeline claim that it will reduce our dependence on foreign oil; those who oppose it claim that much of this oil would be exported. If you read a variety of other articles from different sources, you would uncover additional information about this project and its impacts. You might even find out how many jobs the pipeline project would actually create.

If you are wondering whether this proposed project will harm water resources, you'll see that it takes some work to find the facts. Before you tackle the political and economic issues, you need to understand what the project involves, and how it will affect natural resources.

No matter how complicated or controversial the project, you can look at it as a set of activities for easier evaluation. We start with the same type of approach used in chapter 3, where the review of a project's impacts begins with describing the land and water area disturbed by the project; the natural resources within the area of disturbance; and the full range of project activities.

Analysis of this information tells us how the project affects land and water. The first energy source described here is natural gas obtained by the process of hydraulic fracturing, or hydrofracking. This practice includes a variety of activities and allows us to look at many types of environmental effects shared by other forms of energy development.

Natural Gas: Hydrofracking

Horizontal hydraulic fracturing (hydrofracking) for natural gas (methane) as practiced in 2014 is a relatively new procedure developed in the 1990s for extracting methane from shale deposits thousands of feet underground. First the well is drilled deep into bedrock and extended horizontally through natural gas–containing shale. This increases the area from which gas can be recovered from any one well. The next step is injecting a mixture of freshwater, chemicals, and sand into the well at very high pressure, fracturing the bedrock. The particles of sand prop open these cracks, releasing the gas, which is then recovered along with a portion of the water/chemical mixture. Approximately 30–70 percent of this mixture remains underground. The portion that comes back to the surface picks up additional materials from deep underground and brings them to the surface, notably a large proportion of salt, heavy metals, and naturally occurring radioactive materials (NORM). It takes an average of five million gallons (from 1.5 million to 9 million) of water to "frack" a well; each well can be fracked a number of times. The life span of a well varies from five to thirty years.[3]

How we define hydrofracking influences how we look at its effects. Those who work in the natural gas industry use the term "fracking" in a very specific, technically accurate way—referring exclusively to the underground fracturing process. Based on this definition, "impacts caused by fracking" are limited to only this underground process. However, in order to examine hydrofracking for natural gas comprehensively, we need to expand this definition to include all the activities involved in this process of obtaining energy from natural gas. By this second definition, hydofracking is not limited to the actions that free natural gas trapped deep underground; it also includes the following:

- *Access*: constructing roads, wellheads, staging areas, pipelines
- *Materials*: obtaining and transporting materials used for fracking (sand, water, chemicals)
- *Drilling*: freeing the gas from underground rock formations
- *Collecting*: pressurizing and transporting natural gas (pipelines, compressor stations)
- *Waste disposal*: storing, transporting, and processing waste products
- *Accidents and violations*: spills, leaks, discharges of waste materials

These activities define the area of disturbance—all the land where these activities take place, and the network of interconnected surface and groundwater resources. The "project map" covers a large area, perhaps an entire natural gas deposit or "play"; we can evaluate project activities more easily if we divide this area into watersheds and subbasins and identify their wetlands, streams, lakes, and aquifers.

The following three tables (5.1a, 5.1b, 5.1c) summarize hydrofracking activities that may pollute, deplete, or degrade water resources. These activities are organized into three categories: site preparation and construction; hydrofracking operations; and waste disposal. Once we've examined the effects associated with each activity, we can consolidate this information to describe cumulative impacts from multiple well sites within a watershed.

Table 5.1a covers site preparation and construction, with activities and their effects similar to those described in chapter 3, table 3.1, but with a more extensive area of disturbance because of the large number of sites.

TABLE 5.1A. Hydrofracking: site preparation, construction activities, and examples of impacts

Hydrofracking activities: construction	Examples of impacts
Removal of vegetation	Loss of rare species; degraded, lost, or fragmented habitat; increased soil erosion, stormwater runoff volume and sediment load, invasive species; possible loss of water resource buffers
Grading	Changes in watershed drainage patterns and water supply to streams, ponds, wetlands; soil compaction from heavy equipment; increased erosion and stormwater runoff; loss of (portions of) small wetlands or streams
Construction of access roads, well pads and staging areas, pipelines, compressor stations	Increased soil compaction, erosion, stormwater runoff, impervious surface area; changes in watershed drainage patterns, flooding, and hydrology; disruption of wildlife corridors and migration patterns; right-of-way management using herbicides; habitat fragmentation
Stream and wetland crossings; construction through aquatic buffers	Bank disturbance and increased erosion; changes in water-flow patterns; flow constrictions; floodplain disturbance; interruption of wildlife movement patterns; potential increased flood flows, velocity, and erosion at outflows (e.g., from culverts)
Movement and storage of heavy equipment	Soil compaction; loss of vegetation; increased stormwater runoff and bank and streambed disturbance; floodplain disturbance; water contamination from fuels and operation
Changes in land use: heavy industrial	Altered watershed and ecosystem hydrology and drainage patterns; increased watershed impervious surfaces and contaminant load in stormwater runoff; increased traffic

The construction of each well pad and all supporting infrastructure typically disturbs approximately 8.8 acres. Each area cleared of forest also creates an "edge effect" extending at least three hundred feet into adjacent forest (or other habitat), which opens up the area for invasive plant species.[4] Herbicides used to maintain rights-of-way or other areas may be washed into aquatic and wetland ecosystems and contaminate the water. Construction compacts soil and increases impervious surface, stormwater runoff, and erosion, especially during heavy rainfall and flooding.

Table 5.1b shows a wider range of activities associated with the actual fracking materials, procedures, and operation—mining sand; withdrawing water; handling chemical mixtures; transporting these materials; and collecting, pressurizing, and transporting the natural gas.

Hydrofracking drilling extends thousands of feet belowground to reach shale deposits. Underground pipelines collect the gas via a feeder pipe. Transportation of hydrofracking materials and wastes increases truck traffic and use of heavy equipment, which increases the contaminant load (for example, oil, grease, gasoline) in stormwater runoff from impervious surfaces. Well-pad

TABLE 5.1B. Hydrofracking: materials, transportation, and operation activities and examples of impacts

Hydrofracking activities: materials, transportation, operation	Examples of impacts
Truck traffic (transport of fracking materials)	Increased stormwater runoff contaminant load from roads; air and water contamination from diesel fuels
Mining and transport of silica sand for fracking	Human health effects from silica dust; degraded water resources at sand removal sites
Consumption of water (water removed from water cycle)	Reduced water available for municipal, agricultural, and recreational use, aquatic ecosystems; altered watershed functions, dewatering of small wetlands and streams; lowering of groundwater or aquifer levels
Transportation and handling of chemicals added to water (to aid the hydrofracking process)	Air and water contamination from specific chemicals; human health effects
Flaring (burning off waste gas)	Air pollution from methane gas
Injecting water, sand, and chemical mix into the well and fracturing the shale; use of diesel-powered heavy equipment	Potential groundwater contamination; leaks from failed cement casings; potential migration of contaminants through fractured bedrock; air and water contamination from diesel fuel
Operation and maintenance of compressor stations and pipelines	Air pollution; effects of noise and light on human health and wildlife; movement of pressurized gas via pipelines

access roads bring these contaminants closer to wetlands and streams, increasing chances of pollution.

The chemicals added to fracking water, and their proportions, vary depending on proprietary formulas and individual well and site conditions. A four-million-gallon fracking operation uses about 167 tons of chemicals. If spilled, these chemicals can enter surface waters via runoff from well pads and other sites where they are transported, stored, or handled.

Compressor stations are associated with the emission of air contaminants that are injurious to human health, livestock, and wildlife. Some of these contaminants can move between air and water, polluting both.

Table 5.1c describes a third set of activities associated with the disposal of the wastes generated by hydrofracking operations.

Wastes that result from hydrofracking operations include the following:

1. *Cuttings and drilling mud*: When the well is drilled through rock, the cut rock is removed from the well bore. This may be a large volume of

TABLE 5.1C. Hydrofracking: disposal of wastes and examples of impacts

Hydrofracking activities: disposal of wastes	*Examples of impacts*
Disposal of cuttings and drilling mud	Soil and water contamination from chemicals, salts, metals, and naturally occurring radioactive materials
Recovery, storage, disposal of flowback fluid in on-site pits	Air pollution from volatile toxic chemicals in wastewater from fracked well; concentrated toxic chemicals in waste pits; health effects on wildlife using this wastewater; soil and water pollution from leaks and spills of contaminated wastes into the surrounding environment; human health effects
Disposal of waste: deep-well injection	Potential groundwater contamination; geological instability, earthquakes
Disposal of waste: road deicers and dust suppression	Soil and water contamination (via runoff) from high salt content, heavy metals, naturally occurring radioactive materials; harm to aquatic plants and animals; human health effects
Disposal of waste: wastewater treatment plants or landfills	Water contamination from leaks, and effluent from municipal wastewater treatment plants (not designed to process toxic chemicals, salt, or radioactive materials); harmful effects on aquatic organisms and human health
Disposal of waste: dumping or spraying	Damaged vegetation; soil and water contamination
Fate of the sand/water/chemical mix that remains underground in the well	Potential for groundwater contamination; polluted drinking water

rock, depending on the individual well. Cuttings are brought to the surface by "mud," which is a fluid that cools and lubricates the drill bit. Mud contains a base liquid (water, mineral oil, or synthetic oil-based compound), barium, clay, and stabilizers. It picks up salts, metals, and NORM from the rock it passes through. Cuttings may also contain NORM; often they are disposed of in landfills, or buried on-site. Muds may be disposed of in landfills or deep-well injection pits.

2. *Hydrofracking fluid*: This mixture is injected into the well to keep the fractures in the rock open so the natural gas can be extracted. The fluid contains water, sand, and a mixture of chemical additives—including friction reducers, surfactants, biocides to limit bacteria, and scale inhibitors to prevent corrosion.

3. *Flowback water*: Usually 30–40 percent of the hydrofracking fluid returns to the surface after the hydrofracking operation. It contains the chemical additives as well as materials picked up from the shale underground (salt, heavy metals, arsenic, NORM). Disposal of flowback water includes several options: reuse in fracking operations; transport to industrial treatment plants or municipal treatment plants (which aren't designed to treat many of the contaminants present in this waste); underground deep-well injection (wastes are trucked to Ohio, West Virginia, or other states with suitable geology for these wells). Technology to treat these wastes on-site has not yet been put into practice.

4. *Produced water (includes "brine")*: This fluid is withdrawn from the well as the natural gas is pumped to the surface; it contains high concentrations of the same underground contaminants picked up by flowback water, especially salts. The amount of "produced water" varies by well and site. Methods of disposal include industrial wastewater treatment plants, reuse, and deep-injection disposal wells. Sometimes it is applied to roads as a deicing agent or dust suppressant.

Removal of toxic chemicals from the various hydrofracking wastes remains an unsolved problem not addressed by currently available disposal methods. Each method of disposal listed in Table 5.1c comes with its own set of environmental impacts; these vary depending on the mix of chemicals used in any particular well. Because this information is proprietary, the public frequently does not know which chemicals have been used, so it is tough to evaluate impacts accurately. Where fracking fluids are reused or recycled, the chemicals in them are not removed but become more concentrated with each use, even if the total volume of waste fluid is reduced. Hydrofracking wastes that enter municipal wastewater treatment plants have at times been discharged into surface waters untreated. The hydrofracking "brine"' is washed from road surfaces into soil, surface waters, and groundwater. Deep injection wells pose a high risk for groundwater contamination and have been linked to local earthquake activity in some areas.[5]

Water Consumption

Hydrofracking consumes a large volume of water when each well is fracked. The wastewater it produces cannot be cleaned up and reused as drinking water; it is removed from the water cycle. The extent to which hydrofracking water withdrawals affect water resources depends on how much water the stream or aquifer contains; how much water is withdrawn; competing water-use demands (residential, agricultural, recreational, aquatic ecosystems); and time of year. Large withdrawals from a small subbasin may impair ecosystem and watershed functions, especially during drought. Proportions of water withdrawn from surface or underground sources vary by region. For example, in the Barnett shale region of Texas, 56 percent of hydrofracking water came from groundwater and 43 percent from surface water; in the Susquehanna basin (Pennsylvania), 55 percent came from surface water and 45 percent from local water utilities.[6]

To put the amount of water used for hydrofracking into perspective, using an average 5 million gallons per well, a one-acre pond six feet deep holds about 1.95 million gallons. A one-acre wetland can store 330,000 gallons. An average household of four people uses about 400 gallons of water per day. For a residential development of one hundred houses, that's 40,000 gallons per day or 14.6 million gallons per year.

The nonprofit organization Ceres examined the status of water in states where hydrofracking is in operation. It superimposed hydrofracking well locations on the water-risk maps produced by the World Resources Institute's Aqueduct Water Risk Atlas, finding that many of the locations that support a high density of wells are already severely water-stressed. "Almost half (47 percent) of shale gas and tight oil wells are being developed in regions with high to extremely high water stress. This means that over 80 percent of the annual available water is being withdrawn by municipal, industrial and agricultural users in these regions. Overall 75 percent of wells are located in regions with medium or higher baseline water stress levels. Although water use for hydraulic fracturing is often less than one or two percent of a state's overall use, it can be much higher at the local level, increasing competition for scarce supplies."[7]

Cumulative Impacts

Hydrofracking operations involve multiple well sites, roads, pipelines, and compressors across a large land area and its water resources. Subsequently, the effects from the activities listed in tables 5.1a–c are cumulative. We won't see their full impact from looking at one well-site or compressor station. Table 5.2 is an essential companion to the previous tables because hydrofracking includes multiple well pads, a large area of disturbance, and an extensive infrastructure. The table provides some examples of how the impacts of activities add up, with consequences to ecosystems, watersheds, and human health.

You might customize this table by adding a fourth column that shows the cumulative effect of each of these impacts within a watershed by converting the "per square mile" numbers from column 3 to the area of the watershed.

TABLE 5.2. Examples of cumulative impacts of hydrofracking

Impacts	Unit of measure	Horizontal drilling: 1 well per pad (9 pads per sq. mi.)
Site disturbance (acres per pad plus roads and other infrastructure)	8.8 acres / pad	79.2 acres / sq. mi.
Edge effect for well pads in forested areas (within 300 feet of all disturbed areas)	21.2 acres (of habitat disturbed) per pad (in addition to the 8.8 acres of direct disturbance)	190.8 acres / sq. mi.
Water consumption	4 million gal. per well (NYC-DEP figures); has reached 16–21 million gal. / well in selected wells	36 million gal. / sq.mi.
Chemicals (0.5–2% of fracking fluid)	167 tons per well (from NYC-DEP analysis)	1,503 tons / sq. mi. (USGS figures: 3 million gal. hydrofracking produces 15,000 gal. of chemical waste)
Flowback (typically 30–40% of fracking fluid)	2 million gal. per well	18 million gal. / sq. mi.
Truck trips per well	average 1,200 trips	10,800 trips / sq. mi.

Sources: Ted Fink, *Middlefield Land Use Analysis: Heavy Industry and Oil, Gas or Solution Mining and Drilling* (Rhinebeck, NY: Greenplan Inc., 2011), 31; Nels Johnson et al., *Pennsylvania Energy Impacts Assessment, Report 1: Marcellus Shale Natural Gas and Wind* (Harrisburg, PA: Nature Conservancy, 2010).

An excellent illustrated overview of hydrofracking process and issues can be found in *Hancock and the Marcellus Shale: Visioning the IMPACTs of Natural Gas Extraction along the Upper Delaware,*" from Columbia University.[8]

Accidents and Violations

Hydrofracking activities are beset by accidents, leaks, and spills that can occur when wastes or chemicals are transported, handled, or stored. Storage accidents—leakage from tears in the plastic lining of waste storage pits and storage pit overflows during heavy rainfall or flooding—can contaminate surface or groundwater and aquatic and wetland habitats. In some instances, waste materials have been leaked or intentionally dumped into surface waters or applied to nearby land areas. In some cases, hydrofracking wastewater that was land-applied (sprayed into the air and on vegetation) in adjacent forest (in accordance with the well permit) damaged trees, killed ground vegetation, and changed soil chemistry.[9]

Methane leaks and explosions along pipelines also occur. Cracks or other breakdown of well-casing materials may allow contaminated fracking fluids to leak into groundwater. Demonstrations that "light the water on fire" dramatically indicate the presence of methane in drinking water. Case-by-case information on accidents and violations that have resulted in toxic spills or explosions can be found online.[10] For example, New York State's Hudson Riverkeeper describes over one hundred accidents or violations in its 2010 *Fractured Communities* report.[11]

Based on past records, these incidents most likely will accompany gas drilling (and increase as we drill more wells) because of the extent of proposed drilling, the lack of comprehensive and consistent regulatory oversight and enforcement, the siting of wells in or near sensitive places, mechanical failure, or human error.

Transporting hydrofracking materials and wastes increases the chances of accidents. The New York State Department of Transportation predicts 1.5 million additional heavy trucks on roads annually per seven thousand wells.

- If 1 percent of 1.5 million trucks have accidents, that is 15,000 accidents (per 7,000 wells).
- If flowback waste is spilled in 1 percent of those accidents, that is 150 spills.
- With a potential of 5,500 gallons per spill, that adds up to a potential of 825,000 gallons of spilled toxic wastes per year, per 7,000 wells.[12]

Water Contamination

A 2011 report issued by the US House of Representatives documents that between 2005 and 2009, oil and gas companies used more than 2,500 hydraulic fracturing products containing 750 chemicals and other components. These included 652 different products that contained one or more of 29 chemicals that are

- known or possible human carcinogens;
- regulated under the Safe Drinking Water Act for their risks to human health; or
- listed as hazardous air pollutants under the Clean Air Act.[13]

The chemicals that are used throughout the process of hydrofracking function as friction reducers, corrosion inhibitors, defoamers, emulsifiers, surfactants, acids, pesticides to kill microbes, and gelling agents. These contaminants can follow pathways into water resources via stormwater runoff, belowground leaks (for example, failed well casings), interconnections between aboveground and belowground water, and air pollution (some chemicals like benzene can move between air and water). Harmful effects include the loss of drinking water supply, changes in soil and water chemistry, degradation of aquatic habitats, damage to wildlife and livestock,[14] and a range of human health issues. Consequences of contamination depend on chemical content and exposure. Flaring and leaks from well sites emit methane into the air. Methane also contaminates surface water and groundwater; research has documented the contamination of drinking water by methane associated with hydrofracking.[15]

Information about the chemicals at specific hydrofracking sites is limited, because chemical formulations are "proprietary" information not disclosed by oil and gas companies (unless required by state law). In 2005, the "Halliburton loophole" in the Energy Policy Act provided the industry with exemptions from the Safe Drinking Water Act. The natural-gas-drilling industry is also exempt from provisions of other federal environmental regulations that would require disclosure of chemical information, and from requirements for the treatment, handling, and disposal of hazardous wastes (fracking wastes are not classified as "hazardous" from a regulatory standpoint).

Toxic chemicals can injure humans and other living organisms via three exposure pathways: inhalation (breathing); ingestion (eating or drinking); or skin contact (touching). While we have focused so far on water-based pathways, air pollution can also contaminate water. Many of the chemicals used in hydrofracking operations are water-soluble, which allows them to move into surface water and groundwater. A group of chemicals called volatile organic compounds (VOCs) are flammable and move between air and water. They come from venting and flaring of natural gas, wastewater pits, compressor stations, and engine operation. Examples of VOCs include benzene, toluene, ethylbenzene, and xylene (all present in gasoline, diesel fuel, and coal tar), and formaldehyde. Some of them cause cancer, or are associated with multiple health effects to respiratory, reproductive, and neurological systems. Other chemicals such as lead, chromium, and PCBs are endocrine disruptors.[16] Naturally occurring radioactive materials (NORM) are another group of contaminants associated with gas wells. Levels of radium in wastewater from a group of vertical gas wells in New York State averaged amounts that were more than one thousand times the EPA's maximum contaminant level for drinking water.[17]

A full accounting of these chemicals and their effects on human health and the environment is beyond the scope of this book. Summary sources include a report by the Preventive Medicine and Family Health Committee of the Medical Society of the State of New York[18] and the Endocrine Disruption Exchange (TEDX) in Colorado.[19]

Each state requires gas-drilling companies to test water quality in wells within a specified distance from the drill pad. This distance varies by state; in Ohio it's three hundred feet; in Pennsylvania, one thousand feet. Drilling companies are generally not required to test homeowners' water beyond this area. Within the area or not, homeowners are advised to have their wells tested before and after hydrofracking begins. Testing is the only way to detect many harmful chemicals (such as uranium, radon, strontium, methane, barium) because they do not change the appearance of water. As one water-testing service (WaterCheck) cautions: "Methane has small tell-tale signs . . . water coming from the faucets that spit and sputter or bubbles that appear in the water that do not dissipate over time when left to sit on a counter. Methane levels can build up in your home if it's found in your water. Barium does not have an odor or taste; however, it could make its presence known through strange physical symptoms such as gastrointestinal disturbances or muscle weakness."[20]

Detection of chemical contaminants as they move through water is a challenge. Can regulatory oversight protect us from their harmful effects? An example from Colorado illustrates the difficulty of protection. In December 2012, an industrial plant that processes fracked natural gas near the town of Parachute spilled an estimated 241 barrels of mixed natural gas liquid into the ground, contaminating soil and groundwater. The leak was caused by a faulty pressure gauge on a four-inch pipeline. Though the leak was stopped by early January 2013, the leaked fluids had already washed into Parachute Creek (a tributary of the Colorado River), resulting in high levels of benzene in both the creek and in groundwater. Headlines from the *Denver Post* chronicle the events:

March 28, 2013: "Benzene from gas plant leak polluting groundwater near Parachute Creek"
April 3: "Benzene found in test wells 10 feet from Parachute Creek"
April 18: "Trace amounts of benzene found in Parachute Creek"
April 30: "Health officials say creek pollution controlled"
May 2: "Parachute Creek benzene value exceeds safe federal level"
May 3: "Benzene fluctuates in Parachute Creek, may rise with new treatment plan"
July 18: "Benzene levels in Parachute Creek near gas plant spill double again"

The safety limit for benzene in Colorado drinking water sources is 5 parts per billion. But since the state doesn't define Parachute Creek as a source of drinking water, the creek's limit is 5,300 parts per billion.[21] Less than two miles downstream from the Williams Energy plant, head gates that control the flow of water from Parachute Creek into an irrigation reservoir have been closed since the spill was discovered. According to a May 16, 2013, article in the *Denver Business Journal*, the Colorado Department of Public Health and Environment is "launching into negotiations with the company for a consent order outlining the cleanup of natural gas liquids.... There are no plans at this time for that consent order, expected to be signed within a month, to include fines and penalties on the companies responsible—because the release was not due to negligence but to accidental equipment failure."[22]

These news articles don't provide much information about the effects of benzene contamination on fish and other aquatic organisms, the presence of other contaminants in the leaked fluid, or possible contamination of other surface waters that are hydrologically connected to the contaminated groundwater.

This example from Parachute Creek is not unique; such incidents indicate that existing regulations are not sufficient to protect our water from hydro-fracking-based contamination.

Will hydrofracking chemicals find their way into our water? By reviewing the extensive studies from different states and research organizations that document spills, leaks, accidents, and regulatory violations, you can gauge the extent of this problem. The impacts of hydrofracking will be clearer as additional research information is made available regarding the actual chemicals used at each well and their effects on human and ecosystem health.

Oil

The processes of extracting, processing, and transporting oil can produce significant water contamination. Numerous headlines publicize oil leaks, spills, and related accidents that wreak havoc on land and sea. Oil spilled from drilling rigs, storage tanks, trains, barges, and pipelines finds its way into all of our water resources. Large spills like that from the *Exxon Valdez* in Prince William Sound, Alaska, in March 1989 and the BP Deepwater Horizon oil-rig spill in the Gulf of Mexico in 2010 damaged extensive areas of the marine environment. Land-based spills that ooze into lakes, streams, and wetlands cause similar damage to land and water ecosystems. Additional hazards, including explosions, accompany other accidents—for example, the July 6, 2013, derailment of a train carrying crude oil through Lac-Mégantic, Canada; five tankers exploded, leveling forty buildings, killing forty-seven people, and spilling oil into the town, a nearby lake, and river.[23]

Underground pipelines that carry various forms of oil across the country are grabbing their share of the news. Often these conduits remain unnoticed unless they leak or spill, or unless a new one is being constructed. Earlier in this chapter you had a glimpse of the controversial Keystone XL pipeline. We'll use Keystone as an example in the following overview of pipeline construction and its footprint on the land and then look at the effects from several major pipeline leaks.

An Example from Keystone

The proposed Keystone XL pipeline construction project extends through three states and covers 1,179 miles (distance may vary, depending on route). The pipe is thirty-six inches in diameter, with a life expectancy of about fifty years, and is set within a 110-foot-wide construction right-of-way (sixty feet of this is temporary, fifty feet is permanent). Construction would disturb 16,277 acres; about 10,693 of these would be designated for restoration to previous land use. The remainder would be used for permanent pipeline right-of-way and related facilities like pump stations to maintain pressure and temperature within the pipe to keep the oil moving. Remote-operated valves along the length of the pipeline would be installed at major river crossings, upstream of sensitive water bodies; these valves can shut down a pipeline in the event of a spill or leak. An additional eighteen hundred acres required during pipeline construction includes land for pipe stockpiles, railroad sidings, and contractor yards, additional temporary workspace areas, and temporary housing for pipeline workers. The pipeline would cross 1,904 waterways and disturb more than seven hundred acres of wetlands, riparian areas, and open water.[24]

To describe impacts from such a large project, you need a map that shows its entire area of disturbance. This map includes land required for all the above activities, and the location of all streams, lakes, wetlands, and other water resources by watershed and subbasin. To evaluate a pipeline, consider the following list of project characteristics and activities:

1. *Site disturbance*: removal of vegetation, grading
2. *Construction of facilities*: pipeline right-of-way, access roads, support facilities, pump stations, pipe stockpiles, railroad sidings and contractor yards, temporary workspace areas and housing for workers
3. *Type of oil product and additives*: different products require different temperature, pressure, and other specific conditions to keep the oil moving through the pipeline
4. *Pipeline siting*
   a. Water crossings—arrange all stream, aquifer, wetland, and lake crossings by watershed or subbasin; each crossing should be evaluated separately for connections with groundwater and potential impacts from construction, leaks, or spills
   b. Residential areas—note proximity to residential areas, roads, community buildings, and drinking-water sources
5. *Pipe placement within trench*
   a. Distance to groundwater from bottom of pipeline trench
   b. Distance to surface water from top of pipeline trench (where it passes beneath a water resource)
   c. Pipe size and life expectancy
6. *Reliability and locations of remote shutoff valves*
7. *Revegetation and mitigation plans for disturbed areas*
8. *Containment and cleanup plan in case of leaks or spills*

When you have researched the extent of the pipeline and its related land and water disturbances, you can use this data for evaluating impacts from pipeline accidents.

Pipeline Leaks and Spills

Site disturbance during construction can degrade wetlands, streams, groundwater, and ecosystems. But the main source of environmental damage from pipelines is a leak or spill. To develop a frame of reference for evaluating oil spills, we can take a closer look at a few high-profile pipeline accidents.

Yellowstone River

In July 2011, the rupture of the twenty-year-old, twelve-inch pipeline under the Yellowstone River near Laurel, Wyoming, polluted seventy miles with sixty-three thousand gallons of crude oil, killing fish and wildlife. The immediate cause was attributed to floodwaters that damaged the pipeline and left it exposed; it ruptured under pressure from debris washing downriver. The area was prone to seasonal flooding and erosion; the Yellowstone's riparian zone is shallow and easily flooded. As floodwaters rose in the river, they flowed into this zone. Oil carried by the current settled in these areas as waters receded. Riparian zones support abundant vegetation; the oil coats plants and soil particles and is difficult to clean up. Some sources reported that only 10–15 percent of the oil could realistically be recovered; if that is accurate, the effects of a spill can linger for months or even years.[25]

Mayflower, Arkansas

On March 29, 2013, at least 210,000 gallons (various sources have reported 158,000 to 378,000 gallons, or 5,000 to 12,000 barrels) of tar sands oil (bitumen heavy crude) was spilled from the Pegasus pipeline in Mayflower. By early April, the oil had spread into Lake Conway, about an eighth of a mile from the spill site, and was moving into an outlet stream toward the Arkansas River. Benzene and other chemical contaminants were found in the lake; residents reported fish kills, chemical smells, and symptoms including nausea and headaches. Independent air-testing data showed chemicals found in the samples included benzene, toluene, ethylbenzene, n-hexane, and xylenes. Inhaling both ethylbenzene and benzene can cause cancer and impair reproductive systems, while breathing n-hexane damages the nervous system.[26]

Kalamazoo River, Michigan

In June 26, 2010, a thirty-inch pipeline ruptured, spilling 843,000 gallons of oil into Talmadge Creek and the Kalamazoo River, which flows into Lake

Michigan. Heavy rainfall caused the river to flood, topping existing dams and carrying oil thirty-five miles downstream and into Morrow Lake. The oil-thinning additives evaporated, but the heavy bitumen in the Kalamazoo River sank beneath the surface and traveled along the riverbed. The area of 700-acre Morrow Lake affected by the submerged oil grew from 189 acres in the fall of 2011 to 325 acres the following spring.[27] The cleanup, which includes dredging, cost more than $820 million by 2012 and is expected to increase as dredging proceeds.[28]

Evaluating Oil-Spill Information

Oil-spill information can be organized for evaluation by gathering the answers to a series of questions. Since most oil spills persist in the environment, it's important to consider long-term as well as short-term effects.

1. How much oil has leaked or spilled? What type of oil is involved, and what additives does it contain?
   - Identify total volume of pipeline oil spilled onto land or into water resources (gallons or barrels; one barrel contains forty-two gallons)
   - Describe oil type and additives. Oil is classified into different types by weight. Oil type influences environmental damage and cleanup methods. Crude oil (petroleum) is made up of hydrocarbons and contains small amounts of other compounds (organic materials, metals). Light crude flows at room temperature; heavy crude doesn't flow easily at room temperature and must be mixed with lighter substances to move through a pipeline. Bitumen (or tar sands oil) is the heaviest form of oil and is mixed with additives (for example, benzene and toluene), volatile and often-flammable hydrocarbon solvents and mixtures that dilute the bitumen, producing "dilbit" (diluted bitumen). Bitumen also contains more sulfur, anti-corrosion additives, chloride salts, and sand particles than other types of heavy crude oil. Pipelines moving bitumen operate at high temperatures to keep the oil flowing; the range reported for the Keystone XL is 130 to 150 degrees Fahrenheit. High-temperature pipelines have been linked to a higher frequency of spills because of a higher rate of pipe corrosion.
2. How far has the oil spread?
   - Identify watershed and subbasin(s).
   - Trace potential oil movement routes via interconnected streams, wetlands, lakes, and rivers.

- Map the extent of surface soil and water contamination.
- Identify underground contamination (groundwater or aquifers).
3. What are the short- and long-term consequences? Consider the following:
   - Effectiveness of cleanup (potential long-term contamination of soil and water)
   - Human health effects
   - Changes in aquatic or wetland ecosystems—water quality, water chemistry, habitat degradation, plant and animal health effects
   - Distribution of harmful chemicals throughout a watershed or subbasin
   - Contaminated drinking water
   - Property damage or loss
   - Damage to agricultural resources via contaminated water
   - Loss or disruption of recreational, cultural, and aesthetic resources
   - Cleanup costs, depending on type of oil and methods used

Cleanup

Cleanup plans are more effective if developed before a spill or leak occurs. Local conditions and the type of oil spilled influence containment and cleanup operations.[29] Spills in rapidly moving rivers require cleanup techniques different from those for spills in lakes or ponds. Any spill on land is likely to reach water—whether it seeps into soil and eventually makes its way into groundwater or is washed into surface waters via stormwater runoff or flooding.

Booms (temporary floating barriers for containing an oil spill) are commonly used to block the oil from spreading; absorbent pads soak it up. These fairly standard tools work better in calmer water. In the Yellowstone, at flood flows, methods like booms and pads were mostly ineffective. Flood flows hindered the assessment of spill conditions from boats, contributing to cleanup difficulties.[30]

Skimming (surrounding the oil with skimming vessels so pumps can pull it from the water) and burning (surface oil is towed away from the main slick and burned off) are other methods appropriate only for particular locations. Dredging sediment removes some oil but can spread the contamination farther downstream.

Some methods are not safe for any location, because of their harmful effects on the aquatic environment and human health. Corexit, the dispersant that was used during cleanup of the massive 2010 BP Deepwater Horizon oil spill in the Gulf of Mexico, attaches itself to the oil, breaking it up into droplets and making it "disappear" from sight at the surface and along shorelines. However, Corexit plus oil produces a mix more toxic than oil alone, causing lingering damage to aquatic life, ecosystems, and human communities.

Can oil be totally removed from natural water systems? No one knows how widespread and damaging the full effects of oil spills will be over time. Information about long-term harm from oil spills in the ocean is instructive. In the spring of 2014, the National Academy of Science published a report documenting consequences from the 2010 BP Deepwater Horizon oil spill. The oil caused serious heart defects in developing fish embryos and likely losses for tuna, swordfish, and other species that spawned in oil-contaminated habitats.[31] Twenty-five years after the disastrous 1989 *Exxon Valdez* oil spill in Prince William Sound, the US Geological Survey reports that sea otter populations decimated by the spill have finally recovered to pre-spill levels. Oil persisted in the otters' environment and degraded their feeding habitat; chronic exposure slowed the otters' recovery.[32]

Long-term studies of oil spills from all sources, including pipelines, will continue to expand our knowledge about harm to ecosystems and human health. Oil-pipeline safety depends upon whether or not the oil remains in the pipeline.

Coal: Mountaintop-Removal Mining

Like hydrofracking, mountaintop-removal mining for coal affects large areas, includes an array of activities, produces immediate as well as long-term physical changes to land and water, and illustrates the importance of identifying and describing cumulative impacts on watersheds and ecosystems. Since the mid-1990s, a reduced supply of easily accessible coal has led coal companies to seek more drastic means of obtaining coal in West Virginia, Kentucky, Virginia, and Tennessee. Mountaintop-removal mining for coal is a form of surface mining on the summit of a mountain. Explosives blast away the soil and rock, or "overburden" that lies above coal deposits. Typically the process removes the entire top six hundred to eight hundred feet of the mountain; it may take as much as sixteen tons of overburden to recover one ton of coal. Mining companies may use this overburden to recontour the mountaintop after mining, but they dump much of it, including soil contaminated with toxic mining byproducts, into nearby valleys and streams ("valley fills"). The mining process generates wastes, necessitating the creation of slurry ponds or lakes to hold the byproducts of coal processing.[33]

Mountaintop-removal mining has dramatically changed approximately 1.2 million acres and five hundred mountains in the central Appalachians.[34]

The EPA estimates that by 2012, mountaintop removal destroyed or degraded 11.5 percent of the forests in those four states, and rubble and waste buried more than one thousand miles of streams. By 2012, this form of coal mining eliminated twenty-two hundred square miles of Appalachian forest, with projections of an additional six hundred square miles of land and one thousand streams to succumb in the decade beginning in 2012.[35] How can we begin to describe environmental impacts on such a huge scale?

Area of Disturbance

When you describe a project that affects an entire mountain, it's tempting to overlook some details. The process is easier if you begin by organizing information on maps and tables. As in our previous assessments of projects, the first step is to identify the entire area that will be disturbed by all phases of the mining operation, including the disposal and storage of waste products.

The mining activities that may affect water resources include the following:

- Land clearing; removal of vegetation and soil
- Construction; roads, equipment, and materials storage areas
- Operation of heavy equipment, including trucks
- Blasting to expose coal seams
- Removal of soil and rock above the coal
- Disposal of soil and rock; creation of spoil piles and valley fill
- Coal processing and disposal of coal slurry (containment ponds)
- Grading and revegetating the mined surface
- Accidents, spills, violations

Mountaintop headwaters feed streams that collect runoff and precipitation, carry the water downslope, and join other streams that eventually flow into rivers. So it makes sense to define and evaluate the area of disturbance by watersheds. To assemble this information, you need a map that includes watershed and subbasin boundaries, streams, wetlands, lakes, and other water features, vegetation types, and location of sensitive resources (habitats, wells and other community water sources, recreation areas).

Organizing the Information

To describe water features and mining impacts within a mountaintop-removal area, consider different scales, ranging from large watersheds to individual streams—and the plants and animals that inhabit them. It's one thing to describe changes to a single stream within one fifty-acre parcel; applying that same level of scrutiny to multiple streams in a mountaintop-mined area is challenging because of the size of the disturbed area. For example, the Hobet Mine in West Virginia grew to more than ten thousand acres (fifteen square miles) in twenty-nine years.[36]

Consider all project activities and phases together to evaluate effects on the entire mountain system (including all watersheds). Begin with a summary of mining activities and impacts on water resources in each subbasin within the area of disturbance.

1. Locate the activities within the subbasin and record the extent of disturbed areas (for example, acres of forest cleared, linear disturbance along a stream, wetlands filled in).
2. Describe the physical, chemical, and biological effects of these activities on individual wetland and stream ecosystems within the subbasin.
3. Evaluate effects on overall subbasin health (for example, increased impervious surface, erosion, loss of forest or buffer vegetation, loss of small streams or wetlands).
4. Assess how these effects influence ecosystem and watershed benefits.
5. Combine ecosystem and subbasin impact data for all subbasins within a larger watershed to identify cumulative impacts (including downstream flooding, habitat loss, and water-quality degradation).

You may want to organize this information into a set of tables, similar to tables 5.1a–c. As with hydrofracking, you can divide project activities into the general categories of site preparation and construction; coal removal and processing; waste disposal; and accidents and spills. It's easier to evaluate these activities by subbasin, because of the sheer volume of information and the physical extent of the mining operation as a whole. You can then consolidate this information to evaluate cumulative impacts—both within and adjacent to the area of disturbance, and all downstream reaches of streams and rivers.

Overview of Impacts

Mountaintop-removal coal mining produces complex multiple impacts on watersheds and ecosystems. Overburden dumped into valleys fills in headwaters, wetlands, and small streams. Altered drainage patterns change water storage and flow and influence the amount of water available to sustain aquatic and wetland ecosystems throughout a subbasin. Below valley fills, streamflow may increase, affecting aquatic organisms and raising flood levels downstream.

Mining brings to the surface contaminants and heavy metals that otherwise would have remained underground in coal and rock, including selenium, lead, aluminum, chromium, and manganese. Contaminants are picked up by stormwater runoff or released by valley fills and washed into surface waters. Filtration ponds are constructed to trap mining wastes that contain these chemicals but don't remove them, so overflows from these ponds (during heavy rains or flooding) convey contaminants to adjacent land and water. Each of these contaminants requires scrutiny: How much is released, where does it go, how does it harm natural systems and human health? An increase of contaminants in aquatic ecosystems (for example, zinc, sodium, selenium, sulfate) impairs the health and survival of fish, macroinvertebrates, and other aquatic organisms. Contaminated waters support more pollutant-tolerant species and less diversity. These consequences may be carried far downstream; for example, selenium has been found in coalfield streams below valley fills, with subsequent damage to aquatic organisms and ecosystems.[37]

Cumulative impacts include disruption of natural systems and cycles, habitat loss, decreased biological diversity, increased downstream flooding, contamination of water by mine wastes and by-products, and harm to human health.[38] Extensive habitat fragmentation and "edge" replacement of interior forests changes habitat needed by migratory songbirds and reduces their ability to thrive in the region; some species have declined significantly since this form of mining began. Loss of forests, buffers, small streams, and wetlands impairs watershed health, contributing to the loss of ecosystem and watershed services and benefits—including flood protection, water-quality improvement, drinking water, recreation, education and research, biodiversity, and tourism. Mountaintop-removal mining harms human health via contact with contaminated streams or drinking water and exposure to airborne toxins and dust from mining activities. Studies are showing major health effects on the populations in mined areas, including elevated rates of cancer and birth defects.[39]

Accidents and spills bring additional harmful consequences. For example, in 2000, an impoundment in Martin County, Kentucky, leaked more than three hundred million gallons of coal slurry into an underground mine and then onto hillsides. Regarded by the EPA as one of the worst environmental disasters in the southeastern United States,[40] this leak contaminated residential areas and drinking water and creeks, killing fish and other aquatic life.

In January 2014, a leak from a chemical storage facility (owned by Freedom Industries Inc.) on the Elk River in West Virginia spilled 4-methylcyclohexane methanol (MCHM) and other chemicals.[41] Three hundred thousand residents in nine counties were told their tap water was unsafe for drinking, cooking, washing, or bathing. West Virginia American Water, the state's largest investor-owned water utility, emphasized that once contaminated by MCHM, the water cannot be treated.[42] An article in the *Wall Street Journal* commented on the implications of this event: "The site of a West Virginia chemical spill that contaminated the water supply for 300,000 people operated largely outside government oversight, highlighting gaps in regulations and prompting questions on whether local communities have a firm grasp on potential threats to drinking water."[43] The Elk River spill was one of the most serious incidents of chemical contamination of drinking water in American history.

These mining impacts are occurring in one of the most heavily forested and biologically diverse areas of the United States. Mined-mountaintop reclamation has focused on stabilizing rock formations and controlling erosion; fast-growing nonnative grasses are often planted to quickly revegetate mined sites. The reestablishment of trees is difficult because tree seedlings cannot establish healthy roots and grow well in compacted backfill. Currently there is no evidence that native hardwood forests will be able to recolonize large mined mountaintop sites.[44]

Coal is mined so that it can be burned. Coal-fired power plants produce multiple air pollutants, as well as carbon dioxide emissions that play a significant role in climate change. Some of these pollutants make their way into water. They include

- sulfur oxides, which produce acid rain that changes aquatic ecosystems by lowering pH and altering the water's chemical composition;
- ash and sludge from smokestack scrubbers, which contain toxic metals;
- mercury, which pollutes lakes and rivers and accumulates in fish— which are then unsafe for human consumption.

"Clean coal" is a misnomer, despite industry claims. We have not yet developed and implemented technology to effectively capture and safely dispose of all air and water pollutants from coal-fired power plants, at a cost acceptable to the industry.[45]

Renewable Energy Development

After this brief overview of fossil fuel development, it's tempting to welcome renewable energy development without much scrutiny of environmental impacts. However, any form of energy development may have effects that require mitigation. Even renewable energy can disrupt natural water systems, although in many cases improved project planning, siting, operation, and technology can avoid or reduce environmental damage.

Wind farms sprawl across a large area, with access roads to connect a network of turbines, and infrastructure to convey the energy generated. Species loss and disruption from wind farm construction and operation may have cumulative impacts on local ecosystem functions. Wetlands and small streams require adequate protection from vegetation clearing, invasive species, stormwater runoff, and erosion from roads and turbine sites. Road crossings can interfere with hydrology or wildlife movement between habitats. Herbicides used for right-of-way maintenance can find their way into water. In addition, "although wind energy has many environmental benefits, wind energy development has caused the deaths of birds and bats that collide with turbines and has resulted in indirect impacts to wildlife through behavioral displacement and habitat loss."[46] Many of these negative effects can be remedied by improved turbine design and operations, siting, and project design that take into account the protection of ecosystems and their benefits.

Hydropower requires the use of dams, which dramatically change the flow of rivers, altering aquatic ecosystems and habitat that supports fish. Dam releases can disrupt natural seasonal flow patterns; sometimes dams withhold water and then release it all at once, causing the river downstream to flood suddenly. This disrupts plant and wildlife habitats, changes channel characteristics, erodes banks, and impairs water quality. Sediment in water that enters a reservoir settles to the bottom and may be flushed downstream, depending on water releases and other reservoir management practices. Sedimentation increases turbidity and degrades downstream water quality, harming fish and other aquatic life.

In addition, hydroelectric power plants may significantly affect specific fish populations. For example, large dams in the Columbia River basin have decimated salmon populations; turbine blades at power plants kill young salmon, and dams prevent adult salmon from reaching waters upstream. Some hydroelectric dams now have special side channels or fish ladders to help the fish continue upstream—but even when ladders exist, biologists report significant fish loss at dams along migratory routes.

Mitigation of the impacts of hydropower development on fish involves improvements to overall river and watershed function. For example, in 2011 the National Park Service began the largest dam-removal project in US history—the demolition of two dams that have blocked the Elwha River on the Olympic Peninsula for one hundred years. When completed, the project is expected to open more than seventy miles of habitat to salmon, restoring populations from three thousand to more than three hundred thousand. Dam removal is one large step in a larger Elwha River restoration that includes work to address flooding and restore floodplains, reduce erosion, restore fish stocks, and replant newly exposed reservoir basins upstream of the removed dams.[47]

Sustainable?

Once you've taken a good look at any type of development (residential, agricultural, energy, industrial) and its likely effects on groundwater, surface water, and the watersheds and ecosystems that move water through its cycle, you need to interpret and evaluate that information. Are negative impacts "significant" enough to warrant action? Is mitigation needed? Will mitigation work? And in the larger context: Are these development activities sustainable over time, or will they create substantial cumulative effects in the future? In the next chapter, we'll look at some of the issues that surround evaluation, significant impacts, and mitigation.

part III

CHARTING A COURSE

Working toward Protection

Weighing Significant Impacts, True Costs, and Mitigation

Because any use of groundwater changes the subsurface and surface environment (that is, the water must come from somewhere), the public should determine the trade-off between groundwater use and changes to the environment and set a threshold for what level of change becomes undesirable.

US Geological Survey, *Sustainability of Ground-Water Resources*, 1999

Despite the significance of water in our lives, we cannot protect all water resources from all human impacts. So we must identify the impacts that have the greatest effect on watershed and ecosystem benefits. How can we evaluate impacts from land-use activities to determine which are most significant? The answer to this question depends on community values, as well as scientific concepts.

What is a significant impact? Will mitigation lessen environmental impacts? Will the true costs of development be considered? The answers to these questions will be revealed as we gather information. In the following pages we'll explore approaches for describing significant impacts so we can

- identify locally significant environmental impacts;
- develop local water resource conservation plans;
- use thresholds to evaluate environmental changes; and
- consider mitigation, true costs, and risks.

Environmental regulations usually require that only the most consequential threats (significant impacts) to natural resources receive in-depth scrutiny and mitigation. Faced with an array of potentially harmful environmental effects, someone must decide which are most important. "Significant" is a value judgment—but whose values? What are they based on? Most people don't feel qualified to define a "significant impact"—it is too subjective, site-specific, technical, and apt to lead to controversy.

Lacking criteria to determine local significance of development activities, corporations or individuals proposing a project may rush in to fill this gap. Decisions about natural resources reflect the priorities of those who make this judgment call. A project developer's interest lies in getting the project done as soon as possible, at minimal cost. As a result, a report generated by project proponents or their consultants tends to minimize impacts and favor findings of "no significant impact" even if some important consequences have been overlooked. The community tends to accept the report because specific guidelines for evaluating local significance are not available.

Just as a residential developer knows the minimum number of houses he must build and sell to recoup his investment, citizens must describe limits for negative effects on natural resources. How much environmental change can be allowed as a normal course of doing business—degrading a stream or chipping away at a floodplain—before a bit more is suddenly too much and creates a problem that will be costly to fix? How much water contamination can we allow without losing ecosystem benefits? How much of a watershed subbasin can we change without increasing flood damage? When is one more well or septic system too much? How do we know whether or not harmful effects can be mitigated? While there is no "one-size-fits-all" answer, your community will be better prepared to protect significant resources by developing criteria to address these issues before a controversial project arises.

Identifying limits for allowable harm to the environment is often a matter of context. For example, a watershed subbasin with 50 percent undeveloped land historically contained fifty small wetlands, but half have already been filled to accommodate existing development. The remaining twenty-five wetlands collectively provide services including floodwater storage, water-quality improvement, and habitat. But they fall through the regulatory cracks and are not protected. If you allow one more wetland to be filled, is this loss important? Does it change those collective services? Now consider this: Because of local development pressure over several years, twenty-two more wetlands have been filled, and now only three remain. Is the loss of one of these three significant? What happens when you're down to the last one? At what point does a community protect its best interests by prohibiting a property owner from filling in just one more?

From a landowner's point of view, the municipality has allowed the first twenty-two of these twenty-five wetlands to be filled. Is it fair to prohibit a property owner from filling the twenty-third wetland? Or do you wait until someone wants to fill the last wetland and prohibit that action as a "significant impact"? Where does that leave you with regard to protecting the collective services of the twenty-five wetlands before any were filled? This scenario is oversimplified, but it makes the point that you are better off having a plan for deciding this issue before you lose any of the wetlands. Without a plan, you end up evaluating watershed or wetland services based on an ever-shrinking pool of natural resources. If you keep saying, "We're only losing 10 percent of the wetlands in the county" you are computing that 10 percent based on a total acreage that shrinks each time a wetland is filled. With a plan, you may decide that the wetlands are important for dissipating and storing floodwater, for example. Any activity that interferes with that service would then be considered a significant impact. Or, the wetlands may provide a range of important services within the subbasin, and you may decide to protect all of them.

Regulatory Guidelines

If environmental protection regulations already describe "significant impacts," can't we rely on them to make the "what's important?" decision? Our regulatory system of identifying and mitigating negative changes to the environment may seem adequate—but consider how the system works in real life. Check your state and local regulations for significant-impact guidelines. Keep in mind the following specific questions:

- At what point are there too many buildings in a floodplain?
- How much new construction and impervious surface can a subbasin tolerate before water quality in its streams is significantly degraded?
- How many more wells can your aquifer sustain without diminishing water available for nearby small streams?

State and federal environmental protection regulations are usually not specific enough to answer these questions. Criteria for identifying significant impacts are often associated with large projects and are not likely to apply at a local level—though consequences that are insignificant on a regional scale may be very important locally. While a project that consumes water may minimally affect a state's overall water resources, it can dewater a small subbasin, reduce the water needed by small ecosystems, or deplete a town's limited water supply. A pipeline may appear to have little or no effect regionally, but when an accident occurs, the oil that is spilled may have a huge impact on a local lake, stream, or neighborhood. Significance of an activity that affects the environment is related to not only the size of the activity but also its location, proximity to sensitive resources, and importance of those resources to the local community.

How much guidance do regulations provide? About a third of the states have modeled their environmental legislation on the National Environmental Policy Act (NEPA). According to NEPA, significant impacts should be evaluated in terms of their context and intensity. Context means that an action affects national, regional, and local interests; its significance varies with setting. Intensity refers to the severity of the impact; this includes its effect on public health and safety, proximity to "ecologically critical areas," the degree to which it poses risks to the human environment or affects a threatened or endangered species, and whether the action is related to other actions that are cumulatively significant.[1]

Refer to the NEPA text for the complete guidelines regarding significant impacts. Do they help us answer the three specific questions listed previously? The key phrase in these guidelines is "the degree to which" an action produces a harmful effect—and we need to define "degree" in order to use these criteria to evaluate a real project. NEPA does specify that "significance cannot be avoided by terming an action temporary or by breaking it down into small component parts."[2]

State laws can be more specific. Let's take a quick look at two examples. New York State's Environmental Quality Review Act explains "significance" in terms of the setting, the probability of the impact occurring, how large the impact is and how long it will last, whether or not its effects are reversible, how large an area it covers, and the number of people affected. The New York law lists examples of "indicators of significant impacts on the environment" at the state level for guidance in determining significance.

Here are some samples:

- A substantial adverse change in existing air quality, ground or surface water quality or quantity
- A substantial increase in potential for erosion, flooding, leaching, or drainage problems
- The removal or destruction of large quantities of vegetation or fauna
- Impacts on a significant habitat area
- The creation of a hazard to human health
- Changes in two or more elements of the environment, no one of which has a significant impact on the environment, but when considered together result in a substantial adverse impact on the environment[3]

The California Environmental Quality Act provides similar guidance. Appendix G of the act is an environmental checklist form that provides more specific information for judging "significance."[4] For example, does a project

- substantially deplete groundwater supplies or interfere substantially with groundwater recharge such that there would be a net deficit in aquifer volume or a lowering of the local groundwater-table level (for example, the production rate of preexisting nearby wells would drop to a level that would not support existing land uses or planned uses for which permits have been granted)?
- place housing within a one-hundred-year flood hazard area as mapped on a federal Flood Hazard Boundary or Flood Insurance Rate Map or other flood hazard delineation map?
- place within a one-hundred-year flood-hazard area structures that would impede or redirect flood flows?

These examples from New York and California duplicate the wording and actual phrases used in each law. If we apply the guidelines in these laws to actual projects, we'll discover that first we have to define vague terms such as "substantially," "adverse change," "substantial increase," "significant habitat," "hazard to human health," and "large quantities." Undefined, these nonspecific phrases leave the door open for interpretation—generating inconsistency, controversy, and confusion. So, existing regulations are a starting point, but we must fill in the specifics.

Developing Natural Resource Conservation Plans

By developing a conservation plan, you identify goals for protecting water resources in your locality. A plan helps you evaluate impacts in a larger context—that is, beyond an individual parcel—so that your protection actions make sense. A conservation plan for important natural resources and their benefits can balance business and economic development plans in local environmental reviews.

You can use the information in a conservation plan to develop criteria for impact significance that are understandable, defensible, and locally relevant. The following tips will help you start the process by looking at significant resources as well as the impacts that affect them.

Significant Resources

Identification of important natural features is more than an inventory. While a list of species, wetlands, or streams on a parcel of land is a good start to determine "significance," it's only a beginning. A list of what is present is like taking roll call for an entire school district, without any information about students' age, grade, or class. So if one day an entire grade level or class is absent, how would you know? To be useful, any list, whether of species, watershed basins, or ecosystems, must be accompanied by information about the relationships between these natural features.

Identify the natural resources you want to protect in terms of their extent, value, scarcity, sensitivity to development activities, and ability to recover from harmful environmental changes. As you develop a plan, these are some of the questions you might address:

- Where are the water resources of concern located, and how are they interconnected in the watershed subbasin?
- Why are they important—what benefits do they provide?
- Will changes to an ecosystem or watershed cause you to lose future options for water use?
- Which ecosystems, species, or other features are rare, unusual, or sensitive to change? What are the relevant thresholds for change?
- Is the ecosystem able to sustain itself after it has been changed by development activity? What is its ability to recover, continue to support habitat and species, and maintain services and benefits?

You might identify resources as "significant" (or in need of protection) when they provide important benefits but do not have the ability to recover from harmful activities.

Identify how much of a particular resource has already been lost within a defined area such as a subbasin. Is the remaining resource more valuable because of that loss? Consider ecological value as well as economic or property values, and educational or recreational use. Revisit the benefits described in chapter 1 for additional ideas.

Identify specific ecosystems, habitats, or species that are sensitive to particular activities or that indicate environmental quality through their presence or absence. Sometimes the brunt of proving a project's significant effect falls on one animal species that the state or federal government lists as threatened or endangered. The danger in this is separating the species from the ecosystem that supports it. You can broaden the discussion to include more than just one species by showcasing the ecosystem's importance. For example, the bog turtle is listed as an endangered species in New York State. Any project that affects bog turtle habitat is subject to additional scrutiny. The presence of the turtle indicates specific habitat conditions that also support other species. Contamination by pesticides, agricultural runoff, and industrial discharge negatively affects the bog turtle and its habitat as well as its invertebrate food supply and other species that occupy the same habitat. If bog turtles can no longer survive in a contaminated habitat, what else can't live there? How does this affect the human environment?

Significant Impacts

Perspective influences your approach to evaluating impacts—do you want to limit the size of an activity (for example, new parking lot area) or describe that activity in terms of its effect on a natural resource (for example, the effect of additional impervious surface within a subbasin)? There's a difference between these perspectives that will determine the information you need. Clearing vegetation and grading more than an acre of woodland may be harmful no matter where it occurs; or, the significant effect may come from the position of the impact in relation to a sensitive natural resource, for example clearing and grading any wooded area within a wetland. Is a significant impact building a parking lot for more than fifty cars—or building any paved parking lot (regardless of size) within a floodplain?

Define a level of activity or a project that is significant in terms of its effect on valued resources. Identify the activities or pollutants that threaten the natural systems described in the previous section.

- Does the activity or pollutant disrupt valued ecosystem and watershed services and benefits?
- How does the location or size of the impact-causing activity jeopardize water resources?
- Do specific contaminants persist in the environment?
- What is the cost of losing natural resource benefits? What would be the cost (in dollars or services) of replacing the ecosystem or watershed services that would be lost or impaired? Is replacement possible? Is it expensive? Replacement or restoration is an unlikely option if it's too expensive or difficult to manage.
- Do negative impacts lead to a loss of water resource services that is not reversible (that is, loss of future options and opportunities)?

- Can impacts be mitigated successfully? Or do they result in damage to natural systems that cannot be repaired or restored? For example, clean groundwater is a significant resource that must be protected from contamination. Toxic chemicals that find their way into groundwater can't be cleaned up. Thus toxic chemicals that can contaminate groundwater constitute a significant impact.

Identify land-use activities or contaminants that interfere with the ability of ecosystems or watersheds to sustain themselves in a healthy state. Ultimately, a significant impact damages a natural system to the point where it is no longer sustainable in its original form and is unable to provide its original services and benefits.

Recognize small activities that produce a harmful cumulative effect. If the harm to natural resources is significant, then the activities that cause it are also significant, even if they are individually small. Consider projects that are prone to spills or leaks, and their potential for introducing contaminants that move between surface and groundwater. Look at water quality and water supply for human use as well as for aquatic ecosystems. Identify other stressors on natural systems (for example, drought) that, when combined with development-generated impacts, may produce harmful cumulative effects.

Identify activities that cause impacts that in turn lead to additional impacts. Go through the "scarves in the hat" exercise from chapter 3 to make sure you have pulled up all possible consequences. When you evaluate "significance," consider all effects before concluding which ones are important. To test "significance," answer the "so what?" question: So what if this particular resource is lost? What are the consequences? Are they important?

What is the level of significant impact that would tip the balance and make a project's effects unacceptable? Use information in your conservation plan to develop criteria for environmental impact significance. Criteria that are based on facts are more credible and better able to withstand challenges. You can find additional factual information from examining thresholds that describe risks associated with harmful environmental effects.

Thresholds and Environmental Change

Sometimes environmental impacts are sudden and dramatic—if you dump gasoline into a wetland, the system is instantly contaminated. But environmental changes can also sneak up on you, biding their time as conditions gradually worsen and effects accumulate. You've probably heard the story of the frog in a pot of cold water on the stove. The frog thinks the water is fine as it heats up degree by degree. But eventually the temperature of the water crosses an important upper limit, and the frog is cooked. Some impacts work in a similar manner. The environment may seem to be "all right" for quite some time—until suddenly there's a noticeable consequence that is not acceptable.

Impact "significance" thresholds can help you decide when an activity or its effects become too risky or harmful; they provide a basis for managing or monitoring harmful conditions. Numbers and percentages facilitate describing thresholds, but numbers alone cannot always describe what happens to a watershed or ecosystem when land-use activities pollute water or change drainage patterns. The same impact may have little effect on a resilient ecosystem and a devastating effect on a sensitive system. That being understood, thresholds can be a useful guide for determining when an impact is significant.

Working with thresholds for impact significance is like a medical diagnosis. First the doctor looks at the patient's current state of health, measuring blood pressure and heartbeat and ordering blood tests. These tests produce a set of numbers about the body's chemistry and functions that the doctor must interpret. "Normal" is a range; depending on age, sex, health, and pre-existing medical conditions, the numbers that indicate health or illness vary among individuals.

Similarly, we read and measure ecosystem signs to discover conditions that indicate health; that make the system "sick" so that it fails to function properly; or that may prove fatal without treatment. The conservation plan is the initial checkup, providing a picture of patient condition and ailments; next we interpret numbers and other data, including thresholds, to develop a diagnosis. How does the body, ecosystem, or watershed respond to a sickness, toxic chemical, or change in physical environment?

Often we must deal with gaps in the information available to us. An example is mercury contamination of fish. We do have some measurable limits for how much mercury consumption is "safe" in terms of human health. When mercury concentration crosses human safety thresholds we respond by limiting consumption of fish from certain waters. But what are the limits for harmful effects of mercury on the ecosystem? We don't have clear threshold information for that. And when do concentrations cross a threshold that tells us the risk is too great, and the fish are not safe to eat at all?

Often, crossing a health or safety limit isn't abrupt but is part of a gradual change like the frog's hot water. Indicators, like the plants and animals in an ecosystem, can help us evaluate the condition of water resources and determine when a gradually progressing impact becomes "significant." For example, the Convention on Biological Diversity developed a graphic that shows the scale of land-use changes by the abundance of original species.[5] It begins with a highly natural ecosystem—grassland with 90–100 percent mean abundance of the original plant and animal species. Gradually, the number of species declines as the land is extensively grazed, burned, and finally converted into intensive agriculture, a highly cultivated—but deteriorated—ecosystem, with only about 10 percent of original species present. This is a simplified example, but the concept can apply to any resource that is affected by human use. Where in that continuum from zero to 100 percent does the change become significant? What is the threshold? It's not an easy question to answer, but it's worth discussing as we examine the way we evaluate environmental impacts.

What is an acceptable ecological loss or health risk, and at what point does it become unacceptable? The answer depends on the resources, the nature of the impacts that threaten them, the collective values of your community—and availability of standards that describe the risk to human health, property, and water quality.

Chapter 4 described several environmental threshold measurements or standards; others exist. Threshold numbers can guide you as you evaluate significant impacts, though they do not automatically define significance. Their usefulness depends on their relevance to specific natural resources and locations.

As you develop guidelines or criteria for deciding when an environmental change is "significant," refer to the following examples for evaluating different resources. Footnotes direct you to more-detailed information, samples, and guidance. The Environmental Law Institute's *Conservation Thresholds for Land Use Planners* is an excellent overview.[6]

- *Stream biomonitoring.* You can evaluate water quality by measuring water chemistry and macroinvertebrates on the stream bottom. This method (described in chapters 2 and 4) gives you specific water-quality standards for temperature, dissolved oxygen, turbidity, pH, and other characteristics. It also uses stream-macroinvertebrate data to rate stream condition on a measurable scale.[7]
- *Impervious cover.* The method described by Schueler et al. (chapter 4) predicts stream condition based on the extent of impervious surface in its watershed.[8] Numerical scores correspond to the extent of stream water-quality changes due to surrounding land use.

- *Percent of forested area.* Numerical values for minimum percent of forest cover as an indicator of a healthy watershed can establish limits for impacts. Similar information regarding the extent of forested buffer along the edge of a stream, lake, or wetland also indicates watershed health. There is no single ideal value for percent of forested cover in a watershed or forested buffers along streams, though research described in chapter 2 provides examples.[9] Additional studies are available online.[10]

- *Buffer size.* As described in chapter 2, you can use data that correlate the size of a vegetated buffer with protection of particular services, such as water-quality improvement or habitat, to evaluate land-use effects.[11]

- *Wetlands.* We can evaluate wetlands according to the ecosystem services they provide, as well as their role in the watershed. Thresholds may measure changes in water quality, flood storage, habitat, and overall presence of wetlands as a percentage of subbasin area. Wetland loss represents incremental reduction of a watershed or subbasin's ability to absorb floods and runoff, improve water quality, and sustain habitats.[12]

- *Floodplains and flood data.* Flood-prone areas are vulnerable to subbasin changes that alter the path of water across land and via waterways. Consult the Federal Emergency Management Agency (FEMA) website for floodplain maps[13] and the US Geological Survey (USGS) website for stream gauge and flood data.[14]

- *Water contaminants.* For some of the chemicals that contaminate our waters, you can use the EPA's thresholds, described in chapter 4, keeping in mind their limitations (for example, different standards for enforcement, health, drinking water, and recreation).[15] Consult additional sources for data about specific chemicals.[16]

- *Biodiversity.* Data about the number of species, the presence of sensitive species, and the relative abundance of species types contribute valuable information to developing criteria for impact significance. Approaches include mapping,[17] recovery plans,[18] or methods like the Metropolitan Conservation Alliance's "focal species" analysis.[19]

- *Habitat.* Plants and animals have specific aquatic, wetland, and upland habitat size requirements. State Natural Heritage programs, research centers, and conservation organizations provide information about the habitat needs of particular species. Environment Canada's *How Much Habitat Is Enough?* includes examples for describing habitat thresholds.[20]

- *Well pumping.* Formulas that illustrate the balance between water replenishment and depletion in groundwater resources (described in chapter 4) can calculate thresholds for water use and identify when depletion becomes significant. The USGS is a good source of information,[21] though you may also need local hydrologic data to flesh out thresholds and criteria for significance.
- *Groundwater contamination.* Residential septic system density can exceed a site's ability to treat effluents, depending on soil and water conditions. Data about soil limitations can be used to develop thresholds for development density.[22]
- *Watershed health.* Several measurements, combined, can serve as a basis for rating watershed health—percent of forested cover, percent of wetlands, impervious surfaces, and buffers.

As you develop guidelines for defining significant impacts and risks, a word of caution: before adopting numerical standards, walk through a few real-life examples to discover inconsistencies that you need to address. For example, it's tempting to use 10 percent as the limit for impervious cover in a subbasin. You could designate as "significant" any new development with impervious cover that exceeds this threshold—but according to table 2.1 this includes all residential development with lots of two acres or less. Instead of prohibiting development, you might use a combination of techniques to keep the overall subbasin impervious cover below the threshold—for example, amending site plans to reduce impervious cover, protecting or restoring forested cover elsewhere in the subbasin, and requiring vegetated buffers along edges of all streams and wetlands. That would result in a subbasin with dense development in some areas, and perhaps little or none in other areas, with a total under 10 percent. But there are other points to consider—for example, how close is dense development to a stream or wetland, and how does that proximity influence stormwater runoff and water quality?

Another example underscores the need for caution when using physical size to identify significance, whether of a water resource or an activity that affects it. Laws that protect wetlands may establish a threshold for how large a wetland must be to qualify for protection. But wetland value, whether for habitat or flood storage, isn't necessarily correlated with size. Very small wetlands may be extremely valuable as habitat for rare species. A network of small wetlands scattered across the landscape within a watershed may have an important collective effect on flood prevention and water storage.

Developing local criteria that incorporate specific limits or thresholds for deciding when a particular activity may have a significant impact makes the environmental review process more consistent. Along with criteria or guidelines, include instructions for interpreting them to maximize the chances that every project is subject to the same set standards, evaluated similarly no matter who conducts the review. Developing criteria that define significant environmental impacts is not a perfect process, but it can be very useful—and is more effective than no criteria at all.

When you develop guidelines for evaluating impacts, summarize them clearly and succinctly. The more easily residents can understand them, the better your chances of gaining support for implementing them. You don't need to come up with threshold numbers for every environmental effect. Choose a few measurable key elements that indicate overall water resource health. Develop a scale for measuring the condition of these key elements, so you can monitor, interpret, and display changes due to land-use activities or contamination. These are daunting tasks, but you don't have to reinvent the wheel. Consider an example from the Chesapeake Bay Foundation.

The Scorecard

A watershed (or ecosystem) "scorecard" pulls together significant impact criteria, conservation plan data, and thresholds. It can be a useful tool for identifying and monitoring environmental effects over time and sharing that information with the larger community in a form that is immediately understandable. By placing a numerical value on an environmental condition, a scorecard is a guide for evaluating changes and identifying the activities that push a particular condition like water-quality contamination to a level where it requires remediation.

Chesapeake Bay

The Chesapeake Bay Foundation's annual "State of the Bay" reports present information about key areas of concern that indicate water resource health. The foundation has identified the following critical factors for Chesapeake Bay:

- *Pollution indicators*: nitrogen/phosphorus, dissolved oxygen, water clarity, toxics

- *Habitat indicators*: forested buffers, wetlands, underwater grasses, resource lands
- *Fisheries indicators*: rockfish, oyster, shad, and crabs

The foundation's 2012 report describes the criteria for assigning a numerical value to each of these indicators based on measurable conditions. A sample page is presented below. The foundation scores progress on addressing each of these criteria from one year to the next. This information, which reveals whether the condition of the environment is improving (arrow pointing up),

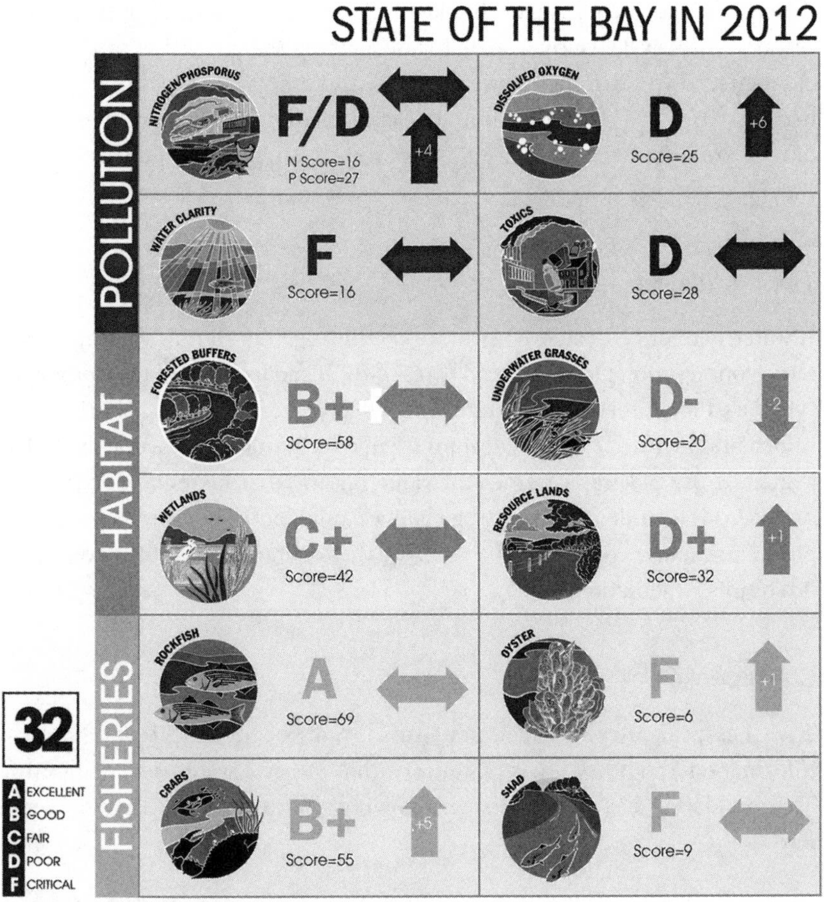

STATE OF THE BAY IN 2012

POLLUTION

NITROGEN/PHOSPORUS
F/D
N Score=16
P Score=27
+4

DISSOLVED OXYGEN
D
Score=25
+6

WATER CLARITY
F
Score=16

TOXICS
D
Score=28

HABITAT

FORESTED BUFFERS
B+
Score=58

UNDERWATER GRASSES
D-
Score=20
-2

WETLANDS
C+
Score=42

RESOURCE LANDS
D+
Score=32
+1

FISHERIES

32

ROCKFISH
A
Score=69

OYSTER
F
Score=6
+1

A EXCELLENT
B GOOD
C FAIR
D POOR
F CRITICAL

CRABS
B+
Score=55
+5

SHAD
F
Score=9

FIGURE 6.1. State of the Bay 2012: Chesapeake Bay scorecard. Courtesy of Chesapeake Bay Foundation, Annapolis, MD.

worsening (arrow pointing down), or staying the same (double-pointed horizontal arrow) informs land-use decisions and environmental-improvement plans. Land-use activities that adversely affect any criteria to a point where the existing grade is lowered are significant to the condition of the bay. Refer to the Chesapeake Bay Foundation website for details and a copy of the annual reports at http://cbf.org/about-the-bay/state-of-the-bay.

Developing Your Own Scorecard

The Chesapeake Bay Foundation and partners begin with a goal that guides their actions: "We have a clear choice: clean water to restore habitat, benefit our children, and create jobs, or delay, resulting in polluted water, human health hazards, and lost jobs—at a huge cost to society."[23] Once you articulate a goal, you can use this scoring approach to identify effects on one wetland, lake, or stream at a time, or on a group of resources within a subbasin or watershed.

While the conditions you measure may differ from the Chesapeake Bay approach, you can apply the same basic method to your circumstances.

1. State your goal; articulate a clear reason for identifying the activities that harm the water resource(s) you want to protect.
2. Identify the "ideal" or "optimal" water resource conditions (for example, prior to development or contamination) that make the resource important to your community. You can base this on ecosystem or watershed services and benefits.
3. Identify several key indicators that describe the condition of your resource. For the Chesapeake Bay, these indicators are grouped into broad categories: pollution, habitat, and fisheries. The indicators within each category are specific measurable elements that are critical to the ecological (and economic) health of the bay. You could use these same categories and adapt them to local conditions, or develop different categories specific to your locality. For a subbasin you may include "flood-control capacity" and monitor the loss of floodplains or flood storage potential (that is, small wetlands and streams, vegetated buffers). Or you may focus on percent of impervious cover within a floodplain or subbasin. Indicators with known effects and thresholds will be easier to evaluate.
4. Develop a numeric system for rating indicators within each category. This can be as simple as a range from optimum to good, passable, poor, and finally unacceptable. The State of the Bay report divides the

"pollution" category into four specific characteristics that reflect the condition of the bay and are relatively easy to measure: nitrogen and phosphorus; dissolved oxygen; water clarity; and toxic contaminants. Specific descriptors and measurements are available for each characteristic, so the rating system is a consistent approach to data interpretation, numerical score, and letter "grade."

5. Compare the current condition of the water resource with the ideal condition. You may also want to measure it against specific known thresholds, like amount of impervious surface, or dissolved oxygen levels required by trout. To display this information, the State of the Bay report assigns 100 as the top grade. Each indicator is given a score, and then the scores are averaged in all categories to determine the overall "state of the resource." The numbers are converted into letter grades for ease of interpretation and communications, from F (below 20) to A+ (70 or better). Everyone knows what an F means. While it may seem too simplistic to reduce the condition of a bay to a letter grade, this system clarifies at a glance the main problems that plague the Chesapeake Bay.

6. Along with each numeric or letter score, indicate whether the condition has improved, worsened, or stayed the same since the last time it was measured. Using arrows, the scorecard shows whether conditions are getting worse or improving from one year to the next.

7. Identify the causes and specific sources of negative scorecard results. This information is critical for planning solutions and improving the indicator scores in the coming year.

You may think that this approach is too complicated for use in your community. It is a matter of scale. Here's an example of how it can be simplified:

- State your goals for the water resource you want to protect (for example, recreation, water quality, habitat).
- Identify ideal conditions and key indicators that measure them.
- Measure current condition and compare results to the ideal.
- Use a simple rating system so that others can readily see effects of impacts on water resource condition.

An impact is significant if it interferes with achieving your goal. On a scorecard, resource decline might look like going from a B to a C. This is a simplified explanation; actual conditions are often complex. But the more specifically you can describe the conditions you're trying to protect, the more meaningful and consistent your designation of "significant" will be.

To manage this rating system, you can choose a few water resource descriptors that indicate environmental conditions—for example, habitats, species, and contaminant levels in water. Your goal might be to protect a stream for trout. Trout require specific water-quality conditions, including low water temperature and a high level of dissolved oxygen. Trout streams support mayflies and stoneflies, aquatic insects that also require these conditions. If stream water quality is gradually declining, you can track changes by measuring dissolved oxygen on a regular basis and assigning it a value. To manage these changes, you have to identify the land-use activity that is causing the dissolved oxygen to decline, so that you can manage the activity's effect on the stream and hopefully reverse the decline. If you identify the source as yard waste dumped into the stream, you can designate that activity as a significant impact on the trout fishery and prohibit or manage it based on this information.

Mitigation

When you evaluate whether or not land-use activity seriously disrupts natural systems, you are also looking at the difficulty and cost of mitigating that impact—eliminating or lessening its negative effects. You might think that mitigation automatically solves the problems caused by harmful activities. Proponents of development projects often assure us that this is so: "Don't worry, we can mitigate all the impacts!" This is an optimistic, can-do statement, and sometimes it's true. But mitigation is also used as an excuse to damage or destroy natural resources without really addressing the damage. The overall promise (making the damage go away) may not hold up to the science of what is required to restore a natural system.

It would be great to believe that across-the-board mitigation of significant impacts takes care of the problems we've been describing in this book. But in reality this rarely occurs. An understanding of mitigation and how it works, or doesn't work, is critical to the evaluation of "significance."

The US Council on Environmental Quality developed the following mitigation definition[24] to guide compliance with the National Environmental Policy Act of 1969 (NEPA):

- Avoiding the impact altogether by not taking a certain action or parts of an action
- Minimizing impacts by limiting the degree or magnitude of the action and its implementation
- Rectifying the impact by repairing, rehabilitating, or restoring the affected environment

- Reducing or eliminating the impact over time by preservation and maintenance operations during the life of the action
- Compensating for the impact by replacing or providing substitute resources or environments

In summary:

Avoid → Minimize → Repair or restore → Reduce over time → Compensate

The rush to use the promise of mitigation to justify an activity that will significantly affect the environment may ignore two important facts. First, the affected resources must be described in terms of how they work and the benefits they provide, so that mitigation can be designed to address these effects specifically. Second, some adverse effects cannot be mitigated. Mitigation is often associated with the statement that impacts will be "minimized to the greatest extent practicable"—a phrase that requires definition. It may refer to convenience or cost rather than to technical feasibility. Mitigation that actually compensates for harm to natural resources can be expensive and require extensive technical planning and design.

Avoidance is the most effective mitigation. Although it may require that site plans are redrawn, or project design is changed, these costs aren't necessarily higher than the cost of trying to replace a damaged ecosystem or its services. Wetlands are a good example. A common misperception is that if you can't avoid destroying a wetland, you can alleviate the damage by simply building a new wetland to replace it. Research demonstrates that this is much more difficult if you want to replicate a natural system that is sustainable over time. You can fill in a wetland quickly and easily, but it's not easy to re-create it as a sustainable marsh or bog ecosystem. Changing a damaging activity or its location to avoid ecosystem damage in the first place may be the most cost-effective and environmentally sound course of action.

If avoidance isn't possible, then NEPA describes other steps. Reducing negative effects on-site (or at least in the same subbasin) is preferable to addressing these effects off-site (in a different subbasin), as it is more likely to protect watershed functions and water supply and quality in local streams, lakes, and wetlands. Ideally, replace a damaged or destroyed ecosystem with the same ecosystem type, including its functions and services. For example, you want to replace a marsh that provides water-quality benefits because marsh plants filter certain pollutants out of the water. Building a pond to "replace" this system isn't valid compensation, because it won't support the same type of plants or provide the same benefits. In fact, claims of "no net loss" of wetlands are misleading because they imply that one type of wetland can replace another type. This ignores differences in ecosystem functions and services.

Restoration

Restoration is a viable option for damaged or degraded ecosystems or watersheds. It is usually more successful than creating ecosystems where none existed before, especially where hydrology is intact. Restoration activities rehabilitate degraded wetlands, re-create natural flow patterns for a stream or river, or use bioengineering to stabilize stream banks with native plants. We tend to dismiss a stream or wetland if it has already been contaminated or degraded, allowing further degradation because it's "already polluted." But often these resources are prime candidates for restoration because they already have the required location with soils or hydrology in place, making it easier to restore ecosystem functions.

Restoration guidelines can be adapted to enhance specific ecosystem or watershed benefits. For example, the US Fish and Wildlife Service offers specific guidance for habitat restoration:[25]

- Return a site to the ecological condition that likely existed before loss or disturbance—for example, removing tile drains or plugging drainage ditches in degraded wetlands, or returning meanders to straightened streams.
- When restoration to its original ecological condition is not feasible, then repair one or more of the original wetland functions, replanting with native species.
- Remove the cause of disturbance or degradation to enable the native system to reestablish itself or become fully functional—for example, fencing livestock out of a stream's riparian area.

Criteria for Evaluating Mitigation Success

The success of any mitigation effort depends on how well it's planned, implemented, and monitored. Standards for evaluating mitigation success should be developed before a project is implemented or approved. First, describe the proposed mitigation:

1. What specific impact is to be mitigated? What activity is the source of this environmental damage?

2. What specific natural resources will the impact affect, and how will it affect them? (Specific means brook trout or bog turtle, not "wildlife"; forested swamp or bog, not "wetland.")
3. Can harmful effects be avoided? If not, why not? Often this question is answered with a no, but a serious attempt should be made to move or redesign project activities or structures to avoid negative consequences. Compare the cost of avoidance with the cost of replacing ecosystem or wetland benefits.
4. If an impact can't be avoided, can its harmful effect on important natural resources be minimized or mitigated? Will changing a land-use activity reduce its negative effects? These changes may include reducing the size of the activity, changing its location in relationship to sensitive resources, or increasing buffers along stream and wetland edges.

Whatever form they take (policy, regulatory, advisory), criteria for evaluating mitigation provide the answers to a second set of questions about how mitigation is designed and implemented.

- *How well does the proposed mitigation plan repair the specific environmental damage produced by the impact? Does it address ecosystem or watershed functions and services?* If you divert water from a small stream by clearing and grading land within a small subbasin, you can't alleviate the damage to that stream by rerouting runoff from the cleared land into a nearby pond. Or if you destroy the storage capacity of five acres of wetland in one subbasin, you won't mitigate the loss to that subbasin by creating five acres of wetland in a different subbasin. You may create some habitat, but you won't compensate for decreasing water storage capacity in the first subbasin.
- *What does successful mitigation look like?* Success produces a healthy system and preserves important functions. Specific watershed or ecosystem characteristics identify success (for example, presence or absence of a particular animal or plant species; adequate stream flow; high-quality water; sustainable ecosystems; low impervious cover; meeting specific environmental thresholds).
- *How will success be measured and monitored, and over what period of time?* Generally, three years is a minimum, but five is better. Identify who will measure, monitor, and decide whether mitigation has in fact been achieved and whether it is sustainable.
- *If mitigation doesn't work during the course of a preestablished time period, who is responsible for fixing the problem?* To ensure

accountability, assign responsibility for maintenance, repair, and, if necessary, replacement of original mitigation practices before the project begins.

Mitigation criteria can be adapted for specific ecosystems. For example, environmental regulations may require mitigation for wetland loss and include a range of options:

- *Creation*: establishing a wetland ecosystem on an upland site (where wetland didn't previously exist) by changing physical, chemical, and biological characteristics of the site
- *Restoration*: reestablishing natural functions in a wetland that has been degraded or damaged
- *Enhancement*: improving specific wetland functions such as water quality improvement, floodwater retention, or habitat
- *Protection*: removing the threat of impacts (for example, conservation easements, land purchase, repair of buffers)

Wetland mitigation must account for specific physical, chemical, and biological features and connections with other water resources on the surface and belowground. What services and benefits does the wetland provide, and which of these are you trying to re-create or restore? How will you duplicate the ecosystem that produces these benefits? If wildlife habitat is one of the benefits, is mitigation tailored to the species of concern? For example, creating habitat for turtles includes not only a wetland pond, but also adjacent upland nesting areas. Wetland creation or restoration projects also include consideration of the condition of the contributing drainage area and the size and quality of the wetland's vegetated buffers. Is a created wetland sustainable? It may look good for the first year but can change drastically after that, depending on invasive species, weather, type of vegetation and soil, hydrology, and the quality of the creation plan. This is why monitoring for several years is critical for long-term success.

You can develop similar mitigation criteria for watersheds. Using a watershed or subbasin map, trace the path of surface water, locate water-storage areas, and describe connections between surface and groundwater. Then you have a basis for evaluating how to compensate for land-use activities that change the system, spread contaminants, or result in too little water in some places, too much in others. How does development affect watershed benefits, and how can mitigation restore them to the system? Adding impervious surfaces within a watershed may be mitigated by the removal of equivalent impervious surface within that same watershed, or by widespread "green infrastructure" practices like rain gardens that capture precipitation and allow it to soak into the ground. Or construction of new impervious surfaces may be limited when the subbasin's total impervious cover approaches thresholds that degrade stream quality.

Many sources describe recommended mitigation practices—for example, the Center for Watershed Protection's "Eight Tools of Watershed Protection,"[26] and the USGS publication *The Practical Streambank Bioengineering Guide*[27] for stabilizing and restoring stream banks.

Whatever the source, a watershed-wide approach is the key to successfully addressing watershed damage and protecting watershed services and benefits.

True Costs

The "true costs" of a project reveal critical information about significant impacts. When a new project comes to town, with promises of jobs and tax revenue, some people are in favor of it right out of the chute. They are eager to support anything that promises more local funds, more jobs, economic revival—which is understandable. But without an evaluation of what a project will actually cost, that acceptance can leave a community vulnerable to misinformation, short-term fixes, promises that are impossible to keep, jobs that aren't sustainable, increased taxes, and in the long run increased costs to taxpayers. These are the costs you often don't hear about until later, when the controversy subsides, construction is completed, and out-of-town workers move on—or when an accident occurs, and the municipality foots the bill for long-term cleanup or restoration of water supply. These costs of development are "externalized"—that is, not paid for by the company building a project but instead deferred to a later date, and to those who live on or near the developed site and who must deal with future problems caused by development activities.

The cost of not protecting water resources is another way of looking at true costs. We often consider ecosystem and watershed services to be free; often we can't see them, so they are not usually factored into decision making and economic evaluations. What does the loss of watershed benefits mean to a community, and what is it worth? What does it cost in future to try to replace natural stormwater-management features, or natural flood-control benefits, habitat, or water-quality-improvement services that water-based ecosystems provide? This too is part of the evaluation of true costs.

In evaluating true costs, we need to look beyond environmental damage on any one site and consider all effects. The following are examples of "true costs" for certain activities:

- *Hydrofracking for natural gas.* True costs, often long-term, include human health impairment; increased medical costs; long-term disposal of contaminated wastes; replacement of contaminated drinking-water supply; cleanup of air, land, or water contamination; watershed damage from sand mining; depleted surface or groundwater supply; sickened livestock and wildlife; habitat damage; and decreased property value.
- *Oil use and transportation.* Leaks, spills, and accidents and their consequences contribute to the true costs of using oil. Typically, environmental damage persists for years after the initial contamination. True costs include negative effects from cleanup chemicals or practices after a spill, leak, or accident; damaged ecosystems; harm to wildlife and human health; and damage to the local economy (property value, recreation, tourism, local business, and agriculture). Contaminant effects on water chemistry can persist for a long time, and though they may not kill everything outright, they can change natural systems, plant growth, and the health of aquatic organisms or wildlife.
- *Residential development.* A residential development may be designed, built, and sold before new residents discover that the water supply is insufficient to sustain the number of houses built, and the community may need to develop new sources. Pet wastes, pharmaceuticals, pesticides, and yard wastes contaminate water, producing cumulative environmental damage. Impervious surfaces from new development can increase flooding and decrease groundwater recharge.

True costs include the loss of specific ecosystem or watershed services. If too many houses are built close together on soils that can't sustain a high density of individual septic systems, those systems may eventually contaminate the area's groundwater. Contamination may spread into lakes, streams, and wetlands. As a result, plants and animals that are sensitive to water pollutants may no longer thrive in the area, people may be sickened, and property values may decline. Project reviewers may not take these costs into account when property is developed, subdivided, and sold. Similarly, construction of too many houses or impervious surfaces within a floodplain reduces the capacity of the natural watershed system to slow flood flows and dissipate them over a larger area. This increases the risk of downstream property damage, and property owners may have to pay more for flood insurance.

Taking a hard look at cumulative impacts on watersheds and ecosystems is an essential step in determining the true costs of development activities. It requires us to dig deeper and take a longer view. All this information factors into whether or not an activity's effect on the environment is considered to be "significant" and whether that effect is something we are willing to risk.

The idea of "risk" is also a value judgment. Even with methods for risk assessment and thresholds for harmful effects, we still have to decide what is acceptable. We don't often think about how much of a gamble we are willing to take as we use chemicals or ignore environmental damage. *Mother Earth News* tells the story of a young boy from a permaculture farmstead in Queensland, Australia, who is given a lollipop in school as a reward. Being raised in a nutrition-aware family, he questions whether it's OK to eat it. His father responds with "What's in it?" After conducting some Internet research about the lollipop's ingredients and how coloring and additives can cause cancer, the dismayed boy decides not to eat it. But concern for his friends leads him to share the information with his teacher and ask her to stop handing out the lollipops, which she does. The principal conducts his own research about harmful effects and officially finds that "the evidence is not conclusive." At this point the boy's mother asks the principal a simple question: "Why would you risk it?" The principal has no answer, but he issues a final position: "We will stop giving lollipops to our students."[28]

This story may seem inconsequential. Certainly eating one lollipop will not harm anyone's health. But the point is not about lollipops, or even feeding our kids healthy foods—it's about how we are so nonchalant about the chemicals we live with, and how we react to the risk they pose to us and to the environment. The question "Why would you risk it?" can help us evaluate whether water contamination is significant. Activities and chemicals that harm health and the environment surround us. We can't possibly track them all. So where do we draw the line for acceptable risk? How do we know what is safe?

Is It Safe?

Evaluating land-development effects on natural water systems can be complex. This chapter provides general guidelines for your consideration. Interpretation of words and phrases, as mentioned at the beginning of the chapter, is key to understanding environmental impacts. Sometimes, all our efforts at evaluating environmental damage can be cast into doubt because of an ambiguous phrase or misleading question that resonates with the uninformed public and that the media can easily repeat.

For example, consider "Is the project safe?" If you answer yes, does that mean there are no significant impacts? If no, is that the same as saying that significant impacts exist? Energy development projects have attracted a lot of attention to this question of safety, and companies reassure the public that "of course it's safe." Natural gas company ads assure us of safety. I submitted a letter to a local newspaper about interpreting the safety issue and received the following response from a reader: "You wrote that 'as currently practiced, hydrofracking cannot be conducted in a manner that is safe' but how do you square your claim with President Obama's Secretary of the Interior, Ken Salazar, who stated that fracking 'can be done safely and has been done safely hundreds of thousands of times.'"

This is a commonly asked type of question, and it's challenging to answer adequately in just a few words. My response included the following points.

- I don't know what personally motivates elected or appointed government officials to take a particular position. I would have to consider their political motivation. Where are they getting their information? Conclusions that come from studies funded by the oil and gas industry are likely to differ from those produced by studies from nonindustry-supported research.
- How do you define a project or activity as "safe"? Is it the same as "protective of human health"? Does "safe" mean the activity won't kill you, or that it won't make you sick? Just what aspect of hydrofracking is safe, for whom, and for how long? The oil and gas industry is spending millions to convince the public that it is safe. But just saying it's safe doesn't make it so. You have to back up that assertion with facts.
- The industry definition of "fracking" is often limited to only the underground procedure of extracting natural gas and doesn't take into account all other associated activities (for example, pipelines and compressors, water consumption, accidents, disposal of contaminated wastes). Are all these activities safe? Are their cumulative effects safe?
- Because of industry exemptions from the federal Safe Drinking Water Act and protective provisions of other federal environmental regulations, it is difficult to assess health effects from chemicals used throughout the hydrofracking process. Nevertheless the medical community documents that chemicals commonly used in hydrofracking include known carcinogens and endocrine disruptors. A rapidly growing body of scientific and medical information documents safety and health hazards associated with hydrofracking; comprehensive public health studies haven't yet been completed.

- The industry states that fracking fluids are mostly water; toxic chemicals used in the process are diluted to the point where they constitute only a fraction (0.5–2 percent) of the fluid mixture. But the industry uses an average of five million gallons of fluid to frack each well, which translates into a large volume of toxic chemicals (for example, 0.5 percent of five million gallons equals twenty-five thousand gallons of chemicals).
- There is no current plan for safely disposing of all the chemical-laced wastes from hydrofracking. They cannot be made to "disappear"—so where do they end up? Underground? In rivers and streams? In landfills? In your water?

Many questions arise over the issue of safety. If we permit an activity, does that mean we agree it is safe? Because if it isn't safe, why would we allow it? This line of thinking leads to still another question: Is the approval of an activity based on the answer to "Is it safe?" or "Do the benefits offset the costs?" In the latter case, who benefits, and who pays the true costs? And how does the question of safety relate to our evaluation of environmental impact "significance" and our acceptance of risk?

In December 2014, a public health review issued by the New York State Department of Health offered the following comments on risk and safety associated with hydrofracking (referred to specifically as "high volume hydraulic fracturing," or HVHF, in the report's text):

> While a guarantee of absolute safety is not possible, an assessment of the risk to public health must be supported by adequate scientific information to determine with confidence that the overall risk is sufficiently low to justify proceeding with HVHF in New York. The current scientific information is insufficient. Furthermore, it is clear from the existing literature and experience that HVHF activity has resulted in environmental impacts that are potentially adverse to public health. Until the science provides sufficient information to determine the level of risk to public health from HVHF and whether the risks can be adequately managed, HVHF should not proceed in New York State.[29]

I leave you with these questions and comments and invite you to investigate. You may discover surprising barriers to the protection of natural resources, beginning with an assurance that "it's safe" or "it won't harm the environment." This investigation brings us to confronting the obstacles to water resource protection, the subject of the next chapter.

7

Overcoming Obstacles
to Water Protection

Everyone is entitled to his own opinion, but not to his own facts.

DANIEL PATRICK MOYNIHAN

Obstacles to natural resource protection can be daunting. We live in communities where people value the environment in different ways. Personal values, opinion, politics, and faulty or incomplete information influence decisions we make about projects and resources. Protecting natural resources is like maneuvering a boat in a stream—the waters ahead are not impossible to navigate, but boulders and other obstacles interrupt progress. You may not be able to remove the biggest boulders, but you can navigate around them.

By anticipating difficulties or barriers, you can address them before they become a problem. The obstacles discussed here fall into several broad categories:

- Lack of data
- Local government issues
- Economic development versus environmental protection
- Property rights and conflicting values
- Burnout

As you consider these challenges, it's critical to keep asking questions. Matt McDonald, in a commencement address at SUNY–Plattsburgh, promoted curiosity and questioning as essential life skills. "Not asking questions is letting someone else think for you," he said. "We all have been and must continue to be question askers."[1]

Asking questions, and then evaluating and interpreting the answers, is essential to ferreting out the facts about environmental impacts. This chapter will help you ask the right questions, provide examples designed to get to the heart of an issue, and elicit helpful answers. You can use information from previous chapters for fact checking, documenting evidence, and assessing cause-and-effect relationships.

Checking the Facts

Many obstacles to effective water resource protection are related to how facts are gathered, presented, and interpreted, and whether they answer the pertinent questions about impacts on specific resources.

Lack of Information about Natural Resources

Whether you are trying to set aside open space or commenting on the effect of a local development project on the water resources in your town, you need facts about ecosystems and watersheds, their services and benefits, and the interconnections between surface water and groundwater. The absence of this data hinders effective protection. For example, site plans that leave out small wetlands or streams impede an evaluation of effects on interconnected water resources. When you don't know where all the wetlands and streams are, you risk siting projects inappropriately, causing headaches for future homeowners who will have to deal with flooded basements or other consequences affecting groundwater, downstream water quality, habitat, or the ability of the watershed to absorb floodwaters.

Environmental reviews tend to divide water resources into different sections like surface water, groundwater, wetlands, streams, water use, and drainage patterns. These divisions hinder the evaluation of impacts on a water system that includes interconnected wetlands, groundwater, and streams—and obscure cumulative effects. You can overcome this obstacle by assembling water information and evaluating it in terms of natural systems.

When you have learned about valuable water resource systems in your community, you will be better prepared to validate your efforts to protect them and to convince others of their importance. Refer to previous chapters for descriptions of specific water resource characteristics. Ideally, a municipality's natural resource management plan includes criteria for identifying locally significant environmental consequences and thresholds for protecting the services provided by wetlands, streams, and watersheds.

Lack of Information about Impacts

Knowing how land-use activities affect water is as important as knowing the resources. In many places, we have become used to an environmental impact assessment routine, where a project applicant presents volumes of data, maps, charts, and numbers to prove that a project won't have any significant effects on the environment. After a process that may be long and drawn out, expensive, and sometimes contentious, a local board approves the project, possibly with minor changes. But the board generally accepts the applicant's findings and does not address impacts unless they've been identified as significant. Proposed mitigation may not alleviate negative consequences.

If a developer presents a detailed site plan at the beginning of the review process, he or she has already invested time and money and is understandably more resistant to changing the plan to avoid negative effects on natural resources. Planning boards may not insist on such changes, because they lack information about how local resources can be affected; they may fear increased project costs, project delay, or the threat of a lawsuit. Applicants and their consultant teams can generate studies that may meet the letter of the law but aren't always based on sound science. Reviewers must consistently check developers' facts, starting early in the process.

We can evaluate development activities proficiently only if we have the right facts at hand—how information is gathered, presented, and evaluated affects the quality of the review. By applying the approaches described in chapter 3 and identifying all project activities early in the process, you can improve both the project review and natural resource protection.

Environmental Reviews

Have you ever heard these statements during a project review?

- "This project won't have any significant effect on natural resources."
- "We have complied with all federal and state regulations, and therefore there will be no impacts."
- "There is no evidence that the impact would be a direct result of this project."
- There will be no adverse environmental effect because "all impacts will be sufficiently mitigated."

These statements may or may not be true; all of them need evidence to back up their claims. Without evidence, they can become barriers to effective protection. The challenge is asking the right questions, checking the facts, and interpreting conclusions. Let's look at the claims one at a time.

"This Project Won't Have Any Significant Impact on Natural Resources"

To discover whether this claim is true, we'll need the answers to a few questions. First, is the area of project disturbance accurate? Have all pre- and post-construction project activities been considered?

Second, to which specific natural resources does the statement refer? Ecosystems? Watersheds? Species? Have these been described in enough detail to evaluate how they're affected by project activities? Have all wetlands and small streams been delineated? What about their subbasins or contributing drainage areas? Were water connections outside parcel boundaries, aboveground and belowground, considered?

Third, have cumulative impacts been evaluated? Do project activities affect downstream systems or groundwater? How do they affect the way ecosystems or watersheds work, or the benefits they provide? Have we pulled all the scarves out of that hat and asked all the "so what?" questions (as described in chapter 3)?

And fourth: How does the statement define "significant," and who decides this? As discussed in the previous chapter, the identification of a significant impact requires defining the term and considering thresholds and standards that make sense locally. By asking these questions and finding the answers, you can either verify that the above statement is true or prove that it's not true (and why) and proceed accordingly.

If you don't already have a natural resources inventory or conservation plan that documents important local environmental characteristics, places, and species, then you have an additional obstacle to overcome as you seek the answers to these questions. This may not be a huge setback, but it can cost you time as you gather the necessary data.

"We Have Complied with All Federal and State Regulations, and Therefore There Will Be No Impacts"

The truth of this claim depends on whether the regulations are designed to protect your local water resources adequately from all the impacts. It assumes that regulations ensure optimum protection for important natural resources. But laws are forged in a political climate. Hopefully, they are based on sound scientific evidence about harmful effects on natural resources, but chances are good that lawmakers worked out compromises so that the regulation could be passed. Those compromises are usually at the expense of environmental protection. Most regulations are limited in their jurisdiction—they protect certain natural resources from some impacts but not all resources from all impacts.

A project that complies with regulations may still be harmful to the environment. Regulations are subject to interpretation. Who determines "compliance"? Does it mean project applicants have met a minimum standard? In other words, could they also comply with the law by exceeding minimum standards for protection of natural resources? Who will enforce "compliance" after the project is completed? Regulations without proper enforcement have no "teeth" to ensure compliance. Know the details about what the law protects, its limitations (what the law does not protect), and how the law is enforced. Sometimes it takes more than one law (federal, state, or local) to protect natural water systems adequately. In fact, that's often the case, since many environmental-protection laws have gaps in their level of protection.

"There Is No Evidence That the Impact Would Be a Direct Result of This Project"

This is a burden-of-proof issue. The claim does not stipulate who looked for the evidence, sources of information, or extent of the evidence search. If you think the evidence exists, your job is to find it. In some cases, you may be able to prove that a project activity causes a negative effect just by looking systematically for the evidence—for example, examining the interconnections between a stream and nearby groundwater withdrawal. For some impacts, establishing proof is more difficult. For example, water contamination can be hard to trace to its source, especially if it comes from nonpoint pollution. In point-source pollution, dye added to water at the source of the suspected contamination highlights interconnections; the dye shows up wherever the water goes. You may need to consult experts to uncover evidence—for example, specific testing for water quality may be required to prove a connection between a project and an affected water resource. Or, you may have to trace water pathways through ecosystems, watersheds, plants, animals, or human communities. You can't establish a connection without evidence. In general, the more evidence you collect, the better your chances to prove a connection between project and impact.

"All Impacts Will Be Sufficiently Mitigated"

This claim is usually accompanied by a reason—for example, because an environmental impact statement (EIS) or a stormwater pollution prevention plan (SWPPP) has been prepared, all impacts will be mitigated. But the fact that an EIS or a SWPPP is available does not mean it has been prepared well or that it describes effective mitigation. These documents may omit pertinent facts, mask impacts, and provide pages of extraneous information. Sometimes their goal is getting a project approved as fast as possible and discouraging comments that lead to delays. Watch out for the phrase "impacts will be mitigated to the greatest extent practicable." This may mean that mitigation decisions are based on convenience or cost to the developer rather than the effective remediation of environmental damage. Mitigation that actually compensates for harm to natural resources can be expensive. This disclaimer allows mitigation that isn't necessarily based on best practices and sound science; so if you see it, request a definition of "practicable" and additional evidence.

To verify whether harmful effects are avoided or reduced, you will need to know whether the review process considered all significant impacts and whether mitigation addresses their consequences. Has the review seriously addressed avoidance as the first and preferred mitigation step? Will the proposed mitigation succeed over the long term? Who is responsible for monitoring mitigation, for what period of time, and what happens if it doesn't work? If there are multiple streams or wetlands, have effects on each of these systems and its benefits been considered? To be credible, mitigation must match impacts. If a project damages fish habitat by contaminating the water, mitigation that improves fish habitat by planting trees on the stream bank doesn't address the contamination issue, though it may improve overall fish habitat by keeping the water cool. Mitigation that addresses injury to aquatic organisms must specify which organisms, since species' needs vary. Once you've established the right connections, figure out how to avoid the impact or lessen its negative results. You may want to contact a water resources professional. If your municipality has adopted mitigation-success criteria as described in chapter 6, it will be easier to evaluate the effectiveness of mitigation measures.

The federal Clean Water Act requires the SWPPP when a developer applies for a National Pollutant Discharge Elimination permit for stormwater discharges on a particular site. The SWPPP identifies the potential sources of stormwater pollution, describes stormwater runoff control measures to reduce or eliminate pollutants and reduce flooding, and describes procedures that will comply with specific permit conditions. A SWPPP can be a useful tool for stormwater management, but it is not intended to substitute for developing effective mitigation of all impacts on water resources.

Getting the Facts Straight

Controversial issues attract confusion about facts. Factions can toss facts and opinions into a single information cauldron, blending them until they are almost indistinguishable. People who represent different sides of an issue bring in their own experts; experts on the same topic but who represent different interests don't always agree. One says drinking water will be contaminated; the other says the water will be fine. Sometimes, local folks caught in the middle are tempted to throw up their hands in frustration and not believe anyone.

"How do I know what to believe?" is a good question; finding the answers may take a little work. First of all, check each expert's references for credentials, experience, and professional reputation. Talk to other experts in the same field and see if they agree. What sources of information does an expert use to back up a position? Look for a variety of credible sources and affiliation with organizations that aren't associated with a particular industry or other development-promoter. Note sources of funding for reports or professional presentations. Does a reputable organization you trust recommend or endorse a particular expert? What interests does that entity have in the outcome of the research? Does a conflict of interest exist?

Confusion about what is "true" or a "fact" leads some people to question science. Our culture currently places a low value on scientific findings. This skepticism can cause real confusion for people seeking the facts about impacts on water resources. During a public meeting in a small town in upstate New York, I was fielding questions about a proposed local water resource protection law. In the midst of my explanation of why the law requires buffers and how buffers protect water quality, a resident who opposed the law jumped up out of his seat, pointed a finger, and challenged me: "How do you know? That's just your opinion! And I don't think it's true." Several others nodded in agreement. "What?" I wanted to say. "Are you kidding?" But of course I thought of a more tactful response. I never expected to have to explain the scientific method and cause and effect—or cite my sources. But I did at that meeting. Challenges to science are increasing—especially when science documents the consequences of environmental degradation and the need for water resource protection.

Education is the key to separating fact from opinion, discerning what is true, and using basic science to evaluate projects and solve problems. Science is not a be-all, end-all solution but does provide a reliable foundation for considering what we risk when we allow water contamination or depletion.

Overcoming this obstacle—confusion about the "facts"—requires relentless education that captures public attention and describes complex issues in simple terms: natural resources provide benefits; if we don't protect resources, we lose the benefits.

Local Government Issues

Local boards and officials make decisions that affect water resource protection. Obstacles to effective protection at this level include lack of information, inadequate funding and support, fear of litigation, lack of political will, pressure to expedite local reviews, and absence of local plans and protection laws.

Inadequate Funding and Support

Compared with the deep pockets of larger corporations, local budgets for water resource protection, evaluation of harmful environmental impacts, and legal consultation are usually pretty thin. The threat of a lawsuit is often enough to panic a local board into approving a project as quickly as possible. Local governments generally lack the funding to hire new staff or consultants to conduct reviews and analyses. In contrast, developers of large projects fund their own consultants and have the resources to present their projects in a favorable light. They also have deep pockets to support legal action when necessary.

There is no single solution for lack of funds. Local environmental-conservation advisory groups, if professional in their approach, may help a municipality save money by collecting and compiling data at little or no cost. Local government can also draw on expertise from universities and local professionals. Grants and other local funding assistance may be available to address a specific problem like flooding, wastewater treatment, or development of local zoning. Citizens' groups can serve local governments well if they provide information from documented sources and establish a relationship of trust with local government. Some states require project developers to fund pertinent local environmental reviews or studies.

Lack of Political Will

The previous methods for overcoming obstacles assume that the local government wants to protect water resources in the best interests of its citizens. But sometimes local boards are set to approve projects that, as drafted, are not in the community's long-term best interests. There are many reasons why this happens, including the political climate (fear of losing an election); fear of litigation; or promises of economic relief, jobs, and increased tax revenue resulting from increased development.

To bolster local political will for environmental conservation, a decision-making body requires public support and evidence. Publicize the facts about (1) specific links between project activities and water resources and (2) effects on ecosystem and watershed benefits important to the municipality. Gather as many residents, organizations, and businesses as possible to support your concern. Conduct an information blitz, distributing the facts to as many people and groups as possible. In your outreach include media, social media, and networks. Make sure the materials you distribute are well researched and presented in easily understood language and an easy-to-read format.

In some cases local officials or boards persist in making choices based on poor information or opinions that are not in the best interests of local residents. When that happens, you may need regulatory enforcement (if existing regulations cover the issue at hand) or legal action. Once a town backs down from substantive water protection to appease a developer who threatens legal action, it sets a precedent that threatens future protection efforts.

Look toward the future and make environmental protection an election issue. The local vote can be a powerful tool for advancing municipal changes—though it may not solve an immediately urgent problem.

Lack of political will to protect the environment is fairly common. In a rural township of New York's Hudson Valley, a development team described, in an environmental impact statement, potential negative effects of a proposed power plant construction project. The local municipality hired a planning firm to review the EIS and evaluate the project's impact on the environment, particularly water resources. The planning team discovered deficiencies in the EIS, including inaccurate wetland delineations, insufficient consideration of watershed impacts, and inadequate evaluation of cumulative effects on water quality from both construction activities and post-construction power plant operation. In response, the municipality (which had already decided to fast-track the project) fired the planning team and replaced it with a consulting firm that was willing to approve the same EIS without requiring any project changes.

Local citizens feel disempowered by this type of scenario, which favors the interests of private developers and industry at the expense of residents' welfare.

Education is one path to reclaiming power over local decisions. Interpreting the data may not be a simple task—it's easier to trust that the "experts" will somehow protect residents' best interests (a risky assumption).

In a different New York township, the applicant for a cell tower construction project submitted an inaccurate wetland delineation. The site contained protected wetlands, but the state stamped the map without field checking wetland boundaries. The town hired a local wetland professional to verify the boundaries, and he discovered that the wetlands were significantly under-delineated (that is, the actual wetlands were more extensive than those drawn on the site map) and extended into the project's area of disturbance. The town went back to the state with a request for another review by wetland staff, and as a result the state changed its official wetland boundary. The tower construction was shifted out of the wetlands.

Local boards with the political will to protect the environment need strong community support—especially when protection requires changing project designs or rejecting certain development activities. Citizens need to encourage planning boards to look at true costs, long-term and cumulative impacts. When local decision-making bodies do not have the political will to protect residents' welfare, citizens may organize to share expenses for legal action.

Expediting Local Reviews

Planning boards are often under pressure to review and approve projects quickly. Their requests for more information may meet comments like: "This project cannot be reduced in size; it would be economically infeasible, and we would no longer be able to pursue it," or "Your [municipality] has taken too long to review this project; it's costing us time and money, and it's time to give it the go-ahead."

It's a good idea to estimate what your municipality will have to pay if water resources are not adequately protected. Some projects are just too big for a particular parcel; sometimes harmful effects can't be mitigated. The argument that changing a development project will make it economically infeasible merits additional scrutiny. There is understandably a bottom line below which a developer will abandon a project if it doesn't return an adequate profit. But often a developer can change a project to protect water resources and preserve the bottom line as well.

Sometimes project reviews can drag on for a long time. This delay can be an obstacle to both development and to resource protection. No matter your position on a particular project and its effect on water, it's in your best interest to support a more efficient review process:

1. Collect relevant data, before a project is imminent, about the area's important water resources, sensitive locations, species of concern, and sources of drinking water and present it to local officials.
2. Decide what constitutes a "significant" impact on the municipality's resources and identify thresholds beyond which you are not willing to go, based on sound science and local values.
3. Provide developers with specific guidelines for development and mitigation, so that they are familiar with the resources of concern before they prepare or review a detailed project plan.

These three actions can save time and money during the review process. In the absence of this up-front information, planning board reviews can continue for months and even years as data appear in bits and pieces. In one case, a developer's team wanted to argue about every environmental-impact detail and insisted repeatedly that the project would not affect the town's water. The facts were not on their side; the municipality was already experiencing problems with water quality and supply in densely developed areas. Every argument that contested the municipality's concern about a potential impact slowed the review process and extended it from one month to the next. If the developer proposing the new project had admitted instead that there would be potentially harmful effects and focused on mitigating them, the review process would have been faster, with less time wasted.

Absence of Local Plans and Protection Laws

Local planning that guides development can improve water resource protection and alleviate problems like reduced property values, insufficient water supply, contaminated water, and increased flooding. Planning tools include watershed management and natural resource conservation plans. Additional details and guidance are available; local plans can adapt recommendations from the EPA and the Center for Watershed Protection (www.cwp.org). The center's "Eight Tools of Watershed Protection" are summarized below and on the EPA website.[2]

- *Land conservation*: Identify important areas to protect within watersheds, including critical habitats, aquatic and riparian corridors, cultural areas, and water supply.

- *Land-use planning*: Address impervious cover as a primary influence on watershed quality; adopt watershed-based zoning.
- *Buffers*: Establish buffer protection along the edges of streams, wetlands, and water bodies.
- *Erosion and sediment control*: Reduce sediment loss during construction.
- *Stormwater best-management practices*: Address groundwater, pollution, and flooding issues.
- *Better site design*: Use improved stormwater management practices that reduce impervious surfaces; promote green infrastructure and low-impact development designs.
- *Non-stormwater discharges*: Identify and control wastewater discharges and other sources of water contamination.
- *Watershed stewardship programs*: Address land-use practices throughout the watershed.

Planning tools can only go so far to ensure actual protection of water resources. Sometimes local water resource protection laws are needed to fill gaps in federal or state regulations (for example, no mandated buffer protection for wetlands and small streams) and address specific community water issues. It is not easy to pass these laws, but consider what their absence allows.

Imagine the following scene. A property owner in a rural residential area that is prone to periodic flooding subdivides a large piece of land. The land contains numerous wet areas and small drainages. Local codes allow this subdivision without requiring professional identification of wetlands and small streams. The state claims jurisdiction for, and maps, a portion of the wetlands on the parcel but doesn't delineate the wetlands outside this jurisdiction. The Army Corps of Engineers, short of staff, cannot send someone to delineate small wetlands unless approached by a municipality with a compelling need. The municipality does not acknowledge a need for this service, especially if it will hinder development, and there are no local guidelines.

Prior to the construction of residences, the county health department approves plans for septic systems. Some of these are located within unmapped wetlands, but the health department asserts that it has avoided the mapped wetlands and that identifying unmapped wetlands is outside its jurisdiction. The project engineer assures new landowners building homes that he can solve all on-site water problems by designing an extensive system of drains and supplying fill, which changes drainage patterns. The extent of federally protected wetlands subject to filling isn't known because no one has delineated them.

The developer clears and grades land and builds driveways. Several old buildings on the site are burned to the ground, contaminating the soil with lead and asbestos. Some of that soil is in mapped and unmapped wetlands and their buffers. The county calls the EPA to monitor removal of all contaminated soil. As the developer continues to prepare the site for subdivision construction, an adjacent landowner notices a significant increase in flooding, stormwater runoff, and erosion on his property. Much of his driveway washes away. Unmapped wetlands on his property extend across the property line into the construction site, where they have been filled. Since the area is already prone to damaging floods during large storms, he is concerned about increased runoff, the quality of his well water (because of new septic system locations and the history of soil contamination), and decreased property value. So he approaches local, state, and federal agencies for assistance.

Each agency responds either by assuring him that it is meeting its regulatory responsibilities, or telling him it has no jurisdiction or staff to investigate his case. The municipality recognizes flooding problems, but political realities obstruct changes in policy or local action. Adjacent property owners watch as all these entities point their fingers at each other and offer little or no help with the problem at hand. The concerns of the neighbor with the washed-out driveway are left hanging.

No one looked at the site's water resources as an interconnected natural system and evaluated effects of development activities on adjacent or downstream property owners and on the people who will buy the homes constructed on these sites.

This is a true story. This situation might not occur in places with more effective regulatory practices, or might be worse in states with less regulatory attention to water resources. Problems concerning water are occurring with more frequency, and so are the storms that increase flooding and disperse contaminants. Local water resource protection laws that are based on sound science and practical action can deal with flooding, water quality, and other issues of local concern that otherwise won't be addressed. Model laws are available through state and federal agencies and watershed-protection organizations. You can adapt them to address specific local conditions or concerns.

Economic Development versus Environmental Protection

National concern about economic development, tax revenues, and jobs parallels warnings about what we lose if we allow our environment to deteriorate. Environmental protection has become a relatively low priority even when the issue is clean water and human health. Federal and state governments cut budgets and staff in the agencies that develop and enforce protective regulations. Exerting significant pressure on government to relax environmental regulations and standards, corporate interests promote the idea that jobs and economic development should have priority over environmental protection. Increasingly, the burden of resource protection falls on local communities— and on the elected officials, planning board members, volunteers, and consultants who participate in local planning, project review, and decision making.

Many people are trying to do the "right thing" for their communities but don't always agree on what that is. Especially in communities that are struggling economically and where people are desperate for relief, it is impossible to ignore a developer's offers of jobs, tax revenues, financial assistance for schools, and other much-needed economic boosts. In these situations, however, it may be prudent to step back and evaluate all the impacts, true costs, and benefits. If we need more information for our evaluation, we get it. If we need time, we take the time. If we are suspicious that a developer's offer sounds too good to be true, we scrutinize it more carefully.

Weighing Economic Development and Natural-Resource Values

Concern about jobs and the economy touches every community in this country, and for good reason. But, unfortunately, this concern often turns into an obstacle to environmental protection efforts, including those relating to water. False statements, including the following, block water protection if they are not challenged:

- Choosing between natural resource protection and the economic health of your community is inevitable; environmental degradation may be the price you have to pay to retain jobs and your standard of living.
- Natural-resource protection means higher taxes.
- Natural-resource protection discourages growth and development.
- Environmental regulations that limit development activities are too costly, discourage development opportunities, and eliminate jobs and tax revenues.

All these statements are weighted in favor of economic development at the expense of the environment. But they can all be countered and disproved with a few key facts about both economics and natural resources. This process may involve a hard look at true costs. How may new jobs, and for whom? Are the jobs temporary or permanent? How will community services like roads, schools, and utilities be affected? Will development require the use of more tax dollars (for increased services, for example) than it brings in? What are the costs of not protecting natural water resources?

Is "hindering development" the same as requiring a developer to change project design or construction activity to prevent the project from polluting your water?

If you can provide this information proactively, instead of defensively, you have a better chance of directing the local conversation. And you can influence those who don't understand the issues or lack access to sound data. Dispelling "environment vs. economy" myths on a large scale will take time, along with the continued and determined efforts of many people who care about water protection.

Developers and local officials often enthusiastically embrace the benefits of development without an accounting of true costs. Examples of these costs include, but are not limited to, the following:

- Long-term loss or degradation of ecological systems and the services and benefits they provide
- Losing or replacing these services over the short and long term
- The cost of accidents and contaminant spills associated with certain types of projects
- Decreased property value (loss or damage) due to flooding, contamination, and other impacts
- Loss of outdoor recreation, tourism, or agricultural opportunities
- Human health costs due to water and air pollution

Another aspect of this "true cost" assessment is the cost of not protecting natural systems and their benefits. This is a relatively new perspective for evaluating impacts and documenting the value of clean air and water and intact native ecosystems. Reports and studies are available, like the Association of State Wetland Managers' *Ecosystem Service Valuation for Wetland Restoration.*[3]

When you consider these true costs, it may be apparent that it's less expensive to protect clean water than to clean it up once it has been polluted—if, in fact, such restoration is even possible. During the 1970s, the Love Canal neighborhood in Niagara Falls, New York, attracted national attention as toxic chemicals from an improperly contained hazardous waste site contaminated groundwater and oozed to the surface. More than nine hundred families were evacuated and compensated for property loss. In 2013, the problem resurfaced; a new group of homeowners on the "rehabilitated" site experienced the same health effects, despite official assurances that the chemicals had been properly "contained" and were no longer a threat.

> "We're stuck here. We want to get out," said 34-year-old Dan Reynolds, adding that he's been plagued by mysterious rashes and other ailments since he moved into the four-bedroom home purchased a decade ago for $39,900.
> His wife, Teresa, said she's had two miscarriages and numerous unexplained cysts.
> . . . She was swayed by assurances that the waste was contained and the area was safe.
> Six families have sued over the past several months. Lawyers familiar with the case say notice has been given that an additional 1,100 claims could be coming.
> The lawsuits . . . contend Love Canal was never properly remediated and dangerous toxins continue to leach onto residents' properties.[4]

Natural-resource protection is critical for sustainable economic development and the long-term health of your community. What happens to the value of your property if your well becomes contaminated? If a municipal well is contaminated, that decrease in value extends to multiple residences. Inadequate watershed protection leads to increased stormwater runoff with a higher contaminant and sediment load. Because of increased turbidity, a local stream, which once attracted tourists and recreational activities, now has water the color of mocha. Would you want to sit and have lunch next to water that looks like the beverage in your mug?

Another statement that obstructs natural resource protection is "The economic need for this project outweighs any [negative environmental] impacts." Ask questions about this statement to clarify its meaning:

- Whose "economic need"? How is the economic need measured, and by what standards? If the need is valid, is this project a good way to address it? How will your community actually benefit from the project and its activities, now and in the future?
- Which impacts? Have you considered long-term and cumulative effects on sensitive resources? What natural-resource benefits will the project affect?
- What does "outweigh" mean? Is economic need more important than the need for clean water? How will harmful impacts influence economic health if they degrade water, impair health, and cause other community

problems? How is "economic need" balanced with loss of ecosystem or watershed services—which we also need? What is the cost of these lost services?

- Have the true ("external") costs of the project been included in a cost-benefit analysis?

Providing these details can counter the notion that environmental protection simply costs too much to be practical. Many of the statements that block environmental (water resources) protection are general conclusions that are not supported by specific facts. They are made to obstruct, not to inform. In response, your job is to inform.

Do Water Resources Have a Dollar Value?

Natural-resources managers all over the world have been concerned about assigning dollar values to our natural assets—species, ecosystem services, and watersheds. Some managers are convinced that we must figure out how to turn natural-resource assets into dollar values to compete with economic interests, capture the attention of the public, and have a chance at protecting those resources. On the other hand, an equally adamant group asserts that, by their very nature, many natural resources cannot be valued in dollars. We have to realize that some things have nonmonetary value—like a breath of fresh air.

You can use both positions. Assigning a dollar value to natural-resource benefits can be useful for some ecosystem services like flood prevention that lend themselves to this type of evaluation. You can calculate the cost of replacing those services or restoring them if they are damaged. On the other hand, attempting to assign a dollar value to ecological services that can't be measured appropriately in those terms risks undervaluing them. You're better off asserting that some valued services cannot be measured in dollars; however, this does not diminish their importance.

Energy Development, Land, and Water Use

The fossil fuels energy industry has become increasingly vocal about imposing development agendas on local communities and pursuing development of public lands. The development of energy sources—even if they are not sustainable, negatively affect our environment (including our water), and don't make sense in terms of true costs to our communities—can be a formidable obstacle to water resource protection. Natural gas or drinking water: Is that a viable choice? Is it cost-effective? Although our national need for energy is a critical issue, we still have to be careful about the tradeoffs.

In response, we can take the following steps:

- Ask questions and find answers about the impacts of energy development—specifically, its true costs—to counter the assumption that energy sources should be exempt from consideration of their environmental consequences.
- Share that information and document it, whether through individual conversations, letters to the media, or other venues.
- Challenge misinformation campaigns with facts.
- Insist on the appropriate independent studies to provide more data about health and environmental effects of energy-development practices.
- Take action, whether through participation in elections, voting, or adopting appropriate legal protection at the local level (for example, restricting or banning oil and gas development that would harm sensitive natural resources).

Most important, we can promote and support the environmentally responsible development of sustainable, renewable energy sources—along with energy conservation—to replace fossil fuels and energy development practices that harm the environment.

Property Rights

Do the following statements sound familiar?

"I have the right to do whatever I want with my land!"
"If you try to make me protect wetlands on my land, it's a 'taking,' and you will have to compensate me for loss of property value!"

Statements like these have aborted discussion, incited controversy, and led to the defeat of local initiatives to protect natural resources. Property rights are in dynamic balance between the common good and individual rights. On one hand is a growing awareness of the need for environmental conservation and protection of community character; on the other is a conviction that property owners should be able to do whatever they want with their land.

Imagine you're at a public meeting to talk about developing a local water resources protection law. Many residents seem to want this to happen, and some are maybe not sure; but they're all concerned about flooding and water contamination. And then someone stands up and says, "You can't tell me what to do on my land. I have the right to do whatever I want on it. You're trying to take away my property rights!" This issue has the potential to derail community efforts to protect significant natural resources, no matter how much information is presented—mainly because people tend to stop dead in their tracks when they hear "property rights" and back off without ever having a real discussion about what it means. Chagrined and silenced, the supporters of new protection initiatives agree: you can't fight that argument. When this happens, the community loses an opportunity for discussion, and residents who support protection lose the opportunity to influence that discussion. The best way to tackle this obstacle is to take charge of the property-rights discussion and expand it.

Examine property rights in light of the benefits that we all derive from natural systems, as well as shared community services (for example, infrastructure, utilities, and roads). The right to protect clean and abundant water is a fundamental property right. What is the value of your property without it? Water doesn't stay within parcel boundaries or county lines. If you own a piece of land, do you also own its water? If the water soaks into the ground or flows off your property, do you still own it? Water is a shared resource that doesn't stay within boundaries. Wetlands, streams, and groundwater are connected. And if what we do on the land affects water, and residents can do whatever they want on their land, then what happens when someone pollutes a shared water supply and contamination moves belowground or downstream to neighboring properties? If what you do on your land contaminates your water, as well as your neighbors' water, do you have the right to deprive someone else of the right to clean water, protection from flooding, and recreational enjoyment? Does your property right trump mine? Dumping contaminants into a stream ruins a neighbor's drinking-water quality. Filling in a wetland causes flooding on neighboring property. Lakefront property is expensive, but what happens to that value when the lake becomes too polluted for swimming?

To some extent we are already constrained from doing "whatever we want" with our land, having agreed as a society that certain activities are detrimental to everyone's health and well-being. We all live in a community that includes environmental resources, or assets. Even if you live in an urban community, your water quality can be affected by what someone else is doing in the watershed that supplies that water.

Sometimes landowners who insist on compensation interpret the protection of water resources as a "taking." This comes from the "takings clause" in the Fifth Amendment of the United States Constitution, which states, "nor shall private property be taken for public use, without just compensation." "Taking" is not a simple issue. However, legal interpretation of a "taking" differs from lay interpretation. "While every regulation of property diminishes the owner's freedom in some respect, not every regulation can be deemed a taking. . . . Interpretation of the Takings Clause, as it applies to regulation of the use of land, is slow to develop and sometimes changes course."[5] While you can consult an attorney for more specific information, you don't need a legal background to urge caution when "taking" is used indiscriminately to apply to any environmental protection regulation.

Conflicting Values

Protection of the New York City watershed has become one of the most frequently cited examples of watershed value and drinking-water supply. Because of stringent protection of watershed lands, wetlands, and streams, and the maintenance of reservoirs in the Catskill Mountains, New York City has been able to enjoy a supply of water that doesn't need expensive filtration. This story has another side, however. Releases of water from reservoirs into the Esopus Creek, historically an iconic trout fishery, have affected the stream ecosystem, changing seasonal water-flow patterns. Significant turbidity has entered the system, to the dismay of recreationists, local residents, and landowners along the creek. Fluctuating water levels and turbidity are associated with increased flooding and with harmful impacts on fisheries, water quality, and recreational value. The issue has become a bitter local controversy.

Reservoirs collect sediment as it settles to the bottom. Over time, this can compromise the ability of the reservoir to consistently deliver clean water and reduces the reservoir's water-holding capacity. Stream-bank erosion along tributaries adds to the turbidity problem. What this means for the future of the New York City water supply is not yet known. Is existing watershed protection insufficient? Will the city's water supply eventually have to be filtered? How will downstream residents deal with a damaged creek ecosystem? The resolution of these issues is not simple. But this situation illustrates how conflicts arise between groups of people that value the same water resource for different reasons. Ironically, even when a watershed is protected to ensure a clean supply of drinking water for one group of people, this same protection may be the cause of water problems for a different group. Both groups face obstacles to water resource protection, with no mutually satisfying resolution in sight at this time.

People look at water resources through different lenses, and as a result they develop different positions based on values. Even in a small community this division can get in the way of effective water-resource protection. For example, a municipality proposing a local water resources protection law holds a series of public meetings. Several of the comments focus on how the proposed law imposes onerous "constraints" on land that otherwise could be available for development. Clearly, this viewpoint sees natural resource protection as an obstacle, or constraint, to development; the opposing viewpoint looks at the same protection as an opportunity to preserve community assets.

This conflict leads to an impasse. The conversation is over, the lines drawn in the sand. When that happens, the discussion must continue, but in a different direction. Without continued conversation, it's hard to make progress toward any agreement. But as with the property rights issue, changing the direction of the discussion, asking questions, using specific examples, and finding any possible common ground is one way to begin. Negotiation does take time, energy, and dedication.

Incivility

Our willingness to talk to each other in a civil manner is critical to the protection of our natural resources. How? Incivility divides people and incites fear and anger—formidable obstacles to getting people together and resolving disagreements. Incivility raises its head at public meetings everywhere.

An illustration of how this can get out of hand began in Nevada in 1999. A controversy arose when local residents opposed the US Fish and Wildlife Service's closing of a washed-out road on Forest Service land. The road was a source of significant erosion in an adjacent creek, which provided habitat for a dwindling population of protected bull trout. Vehement opposition to the road closing grew into a larger movement called the "shovel rebellion," which spread throughout a larger area and into neighboring states.

This dispute degenerated into violent personal attacks. For example, two county commissioners vying for floor time at a public meeting had to be physically separated by the sheriff; the former publisher of a local paper expressed his position by shooting an officer's dog in the middle of town.[6] Local residents harassed and threatened Forest Service employees and their families, prompting one Forest Service supervisor to resign because of concerns for the personal safety of her family and her employees. In January 2000, that official,

Gloria Flora, took her plea for civil discourse to other western towns, including Hamilton, Montana, where I lived at the time. The setting resembled a scene from an old western movie. As the crowd of nearly three hundred crammed into the largest meeting room in town, pickup trucks from Nevada pulled up, unloading out-of-town folks who looked like they meant nasty business. I didn't see any shovels or guns from my seat in the middle of the row, but as the air temperature rose I considered escape routes and developed a plan for what to do if someone started shooting.

Ms. Flora was heavily guarded during her entrance and exit from the building. Her talk was an eloquent plea for civility among people—even when they don't agree, have a grievance, or are angry. The event was fringed by police officers and was tightly controlled. Luckily, there were only a few angry outbursts, quickly quieted, and no shooting. But I never forgot the feeling of "not safe" at a public meeting. Nor did I forget the power of that call to civility.

Most of us will never experience such an extreme—but even a small dose of name-calling, threats, or nasty rhetoric in a small local setting can do a lot of damage. How do you deal with incivility that sidetracks, intimidates, or demeans you or your position? The response requires awareness, a cool head, and a tough approach. It's up to all of us to insist on civility from everyone in public settings and allow incivility zero tolerance every time it raises its ugly head.

It may be tempting to be uncivil to someone who is untruthful, disrespectful, or otherwise attacks you or your position. Passion for protecting the environment is an asset that can energize community members. But it can become an obstacle if you allow yourself to escalate the nastiness, by falling into the trap of being uncivil or being drawn into heated argument in a spur-of-the-moment reaction. The next chapter will discuss additional tips for dealing with confrontation.

Burnout

Merriam-Webster defines burnout as "the condition of someone who has become very physically and emotionally tired after doing a difficult job for a long time." Does this describe your experience? Burnout is a tough obstacle to overcome because it's so personal. Those who fight for environmental protection will be continually challenged by others who insist that their own practices are safe or necessary, don't believe that the health of the environment is threatened, or don't have access to sound information. To prevail, you need to be dedicated, informed, and involved for the long haul. Environmental problems are rarely solved overnight, and personal dedication to a cause can indeed lead to burnout.

To counter this, you need to regularly replenish your personal energy. Get outside and enjoy the natural environment, as a reminder of why you are dedicating your time to protecting it. Experience the pleasure of water, whether a fountain in a pocket park or a swimming hole in a local stream. Talk with a few like-minded people, and share appreciation of the natural resources you value.

Some of the obstacles to progress are formidable. You may feel like you're trying to light a whole room with one candle. If you join with others who share your goals and concerns, collectively you'll shed more light on important issues. Don't expect to ignite all the unlit candles around you at once; it's more likely to be a "one-at-a-time" task. Collaborate with others to develop a strategy, and put it into practice. This will give you a foundation for action and a consistent approach, increasing your chances of success.

8

Strategies for Action

[The protection of oceans] is the work of nations, but such goals require pressure from ordinary citizens if there is to be any hope of bringing them about in the face of opposing political and economic interests.

New York Times editorial, July 15, 2011

Environmental protection requires a strategy for action. Corporations develop business plans and marketing strategies to give their activities context and focus. Municipalities draft comprehensive plans and economic development plans. Developers employ public relations strategies to promote their projects to municipalities, government agencies, and the public. Similarly, achieving lasting protection for water resources requires clear goals and a strategy for achieving them. That strategy includes presenting and marketing environmental protection to decision makers and those who influence them.

So far, we've looked at natural systems and the human activities that affect them. We set the stage, and now we move into the action. This is your opportunity to become a part of the story, influence its progress, and affect the outcome.

What's in a Strategy?

A strategy or plan of action shapes your approach to a project or a problem depending on what you want to accomplish. In this chapter we'll look at some of your options for individual or group actions and how to increase effectiveness and maximize the chances that your actions will be successful.

Overcoming the obstacles described in chapter 7 requires a plan of action. Consider this scenario: A news story reports that a proposed project in your community will create five hundred jobs and enrich local tax coffers by hundreds of thousands of dollars. And it won't harm the environment! You suspect that some of these claims are inaccurate or even untrue, but before you can take a breath to request more details, an uninformed public enthusiastically embraces the project. So now you are not only questioning the initial claims but also countering your neighbors' position.

What can you do? First, become informed: investigate the statements, ask questions, evaluate the proposal, and consult the original source of the facts. Maybe some claims in the news story aren't true. Or maybe they are true, but other important facts have been left out, such as costs to the environment and effects on local water resources. In either case, crafting a strategy helps you make the best use of your time and effort and increases your chances of success.

The type of strategy depends on the issue. Responding to a development proposal requires a strategy that differs from the strategy you'd use to initiate a proactive project such as a conservation easement or a watershed management plan. Since even the most innocuous-seeming protection plans can face vehement opposition, be prepared by anticipating and planning for responses to possible obstacles.

A few tips to get started:

1. *Decide what you want to accomplish.* What would "success" look like? If the ultimate goal seems too big, too far in the future, too formidable, you may need to break it down into more manageable pieces.

2. *Be specific about your position.* "Opposing development" is not specific enough. Describe the problem and what you want as a result of your actions.

3. *Keep your approach positive.* Focus on what you want rather than what you don't want. It's more productive to start with what you want to protect and why, and how the activities you oppose will affect it. It's also easier to attract supporters with a positive plan of action.

4. *Do the research.* Check the facts, and make sure you get them right. For a viable case, the facts should support your goals.

5. *Identify sources of support.* Contact individuals or groups likely to support your position, project, or idea.

6. *Pinpoint your targets.* Who has the authority to make the decision you want to influence? Who provides input to the decision? What do you want that person or group to do, and what will they need in order to do it?

7. *Establish a timeline.* Identify all regulatory, administrative, or other deadlines for permits, reviews, review periods, public hearings, or other decision points.

Are you developing a strategy before a problem arises, or are you reacting to a specific project or issue? Sometimes you can develop a stronger position if you include both—for example, providing political support for a local ban on hydrofracking and also supporting new renewable energy development. Or, identify the water resources you value in your town and propose a local ordinance to protect them from activities that cause significant impacts.

Building Community

Whatever your project, it's helpful to consider what you value about water in your community. Responses can bring people together as they relate to a shared positive value. Building community is a positive force for environmental protection. In the small town of Narrowsburg, on the Delaware River, residents of all ages and backgrounds worked on a project to construct an eagle-viewing platform on Main Street. Flanked by barrels brimming with colorful flowers, the platform contains a field scope, a couple of benches, and an attractive US Fish and Wildlife Service display panel about viewing bald eagles along the Delaware. This mini-park has become a community gathering place focused on the river and its eagles. Similarly, whether you're in Roscoe, New York, or Saratoga, Wyoming, it's hard to ignore the community focus on healthy rivers and trout fisheries—resources that draw fishers from all over the country, shape community identity and appearance, and bolster the local economy.

When people focus on the value of a shared resource, they find common ground for working together. The Planning Department in Larimer County, Colorado, appointed a local advisory group to review land-use issues within the rural community of Laporte. One of the priority issues in Laporte was renovation of degraded, unimproved area along the Poudre River. The advisory group, representing a cross-section of the community, included local old-timers ("we like this place the way it is") as well as newcomers ("with some improvements this community could be really great"). People who had lived in the area all their lives didn't trust newcomers, many from California. At first there was so much suspicion between these groups that the chair intentionally adopted a strategy for getting them to work together. He asked them all, "What do you want your community to look like in twenty years?" In the ensuing discussion the group members discovered that they had more in common than they thought and that they could agree on a few priority issues—like establishing a community park along the river. All members— Republican and Democrat, old-timers and newcomers—agreed that they weren't on the right path to achieve that goal without some changes in strategy. County approval of the riverbank renovation and park project took several years, but the group persisted and convinced the county that it was a priority. The Laporte Planning Advisory Committee then found funding and built the park. Today, there's a beautiful community park on the Poudre River, with stabilized banks, picnic areas, and shade trees.

Changing the status quo often begins with getting people together.

Coalitions

Coalitions bring together individuals or groups for a common purpose. They elevate individual lone voices, with little power, to a group voice—a great advantage in drawing attention to an issue, gaining support, and achieving success. A dramatic example comes from the town of Dryden (population 14,500) in central New York, where natural-gas-industry representatives approached individual landowners to sign gas leases. One by one, pressured to sign, concerned residents found each other and shared their stories, forming a coalition that promoted a town-wide ban on hydrofracking. Residents supported the ban by a 3–1 margin, and the town board cast a unanimous bipartisan vote.

The gas industry sued Dryden; in 2012, a state court ruled in favor of the town. In May 2013, an appellate court ruled unanimously in favor of the town. And in June 2014, a third round of litigation ended with the state's highest court ruling that towns can use local zoning laws to ban heavy industry, including oil and gas production within municipal borders. "The people of Dryden stood up to defend their way of life against the oil and gas industry. And, against stiff odds, they won," said Deborah Goldberg, Earthjustice attorney.[1] "Today the Court stood with the people of Dryden and the people of New York to protect their right to self-determination. It is clear that people, not corporations, have the right to decide how their community develops," said Dryden deputy supervisor Jason Leifer. "Today's ruling shows all of America that a committed group of citizens and public officials can stand together against fearful odds and successfully defend their homes, their way of life and the environment against those who would harm them all in the name of profit."[2] Each lone individual who experienced something that he or she felt was just not right, and shared that conviction with neighbors, had a role in this blazing success story.

Another success comes from the state of Maine, where residents formed the Protect South Portland coalition. Motivated by concern about potential oil spills, difficulty of cleanup, and air pollution, the group worked for passage of a Clear Skies Ordinance to prohibit loading tar sands oil and other crude onto tanker ships at the city's port. Despite significant spending and lobbying by the oil industry, in July 2014 the City Council voted 6–1 to pass the ordinance, in the presence of hundreds of cheering volunteers. A representative from the Natural Resources Council of Maine said that if the vote is challenged in court, South Portland plans to defend its ordinance and enlist support from other environmental groups and supporters via a national online campaign.[3]

The significance of these, and other, success stories from local coalitions spreads beyond Dryden or South Portland to other communities, giving citizens a boost and motivating them to act collectively to protect their local resources. These stories do not always involve legal or government action; land trusts are an example of people coming together to protect valued resources by promoting conservation easements in their communities. The national Land Trust Alliance documents more than seventeen hundred land trust organizations across the United States. Each of these has local volunteers and supporters who network with their neighbors, encourage easements, and work to make conservation a reality in their communities.[4]

Coalitions offer neighbors an opportunity to share tasks and expenses and draw upon each other's talents and expertise. When one person is burned out or discouraged, others can keep the effort going and help energize the weary. Coalitions serve a purpose, whether the goal is to protect natural resources before they are threatened or to counter a direct and immediate threat—though people may be more likely to become involved when they're motivated by fear or anger.

Common Goals

Coalitions work. But their success is not accidental. A coalition's effectiveness begins with a clear set of goals and ground rules. The Chesapeake Bay Foundation (CBF) states its goal as "Saving a National Treasure"—its mission statement summarizes the foundation's plan: "Saving the Bay through education, advocacy, litigation, and restoration." If you are considering joining CBF, you know its goal right away and have an idea of how you may participate or contribute.

Individuals forming a group reach a mutual understanding: no matter your other differences, you agree to help each other achieve a common goal. Developing a written mission statement, even an informal one, reminds group members of their purpose. You agree to leave issues alone if they are not related to achieving common goals. This can be a challenge, but personal disagreements within a coalition can, if left unchecked, damage the group's cohesion to the point where it loses focus and can't accomplish its goal. Group members may need a mediated opportunity for venting to clear the air, but only within the "safe" context of the group, and not in public.

Many effective coalitions have started around a kitchen table, where a few friends get together to discuss an issue. Whether it's a gathering of neighbors, or a coalition of strangers with a website, newsletter, and budget, pay attention to the group voice you create. It's often a good idea to establish some basic ground rules right at the beginning.

Credibility

Maintaining credibility is at the top of the list. Your first task is to become informed and reach out to others with the results of your research. Accuracy establishes credibility. Always base actions and outreach on well-documented information. The individuals entrusted by a group to attend a meeting or send letters to the press should be civil, present facts, and not overstate the case. Public testimony is not the time for inflamed rhetoric or accusations. Remember, you are in the business of collecting and sharing the facts that support your position. The last thing you want is an uninformed letter to the editor that gives your new coalition a negative reputation, which is hard to reverse.

Credibility is important every step of the way. Once you lose it, it's hard to recover. When achieving your goal depends on convincing a skeptic, any lapse in credibility can be fatal to your cause. The larger the coalition, the tougher it is to control letters to the editor, position statements, or press interviews. A group requires leadership and a division of tasks among members. Effective leaders understand the goal, issues, and ground rules; guide the group to follow its mission; run meetings effectively; and stay organized and on task.

One on One

Coalitions are built one person at a time. As essential as the group voice can be, it is the result of many one-on-one conversations. Extending these conversations to local officials and other decision makers establishes common ground and builds your reputation as individuals in a group who respect local officials' positions, fears, or concerns about the issue at hand. Bring in people who can help you, and identify the elected officials who will stand up for the community's best interests regarding water resource protection.

During conversations to recruit members for your group, be prepared to answer their questions: What do you want me to do? How much time will I need to spend on it? What's the urgency for that action? What difference will it make? While it's important for your coalition to hear different viewpoints in the community, it's not realistic to expect everyone to join. There will always be people you can't persuade and who may work against your purposes. Who harbors a grudge, has a personal ax to grind, or benefits from a particular outcome that does not support your goal? Conserve your energy by encouraging potential allies who want to make a positive contribution rather than trying to persuade immovable opponents who may try to slow your progress. Persevere, hear them out, and flow around them.

Networking is an outgrowth of individual conversations that can build community. When a group expands connections to others, its energy base and area of influence also increase. Networking seems obvious, but we may neglect it until we encounter an issue that spurs us to action. Networking is a great tool for developing protective plans, finding information, gaining support and encouragement, and learning about how others have tackled similar problems and challenges. Many agencies, nonprofits, universities, research institutions, and volunteer organizations across the country can help you with information, inspiration, and strategies. Often, if you start with one or two groups you know and trust, they will provide links to others.

Social networking presents endless possibilities for reaching out to other individuals or organizations across the country that can share their expertise and offer support.

Funding and Group Identity

When a group's work requires conducting scientific studies, retaining legal expertise, or paying for publicity, fund-raising becomes a necessity. Raising money isn't easy, but it can foster a shared effort that unites people. Be imaginative in generating ways to raise funds. Groups can sponsor an array of creative fund-raising activities—from hosting events to providing services.

Clear Skies Over Orangeville (CSOO) advocates responsible energy development in Wyoming County, New York. The group was organized because local government supported the siting of a proposed wind farm in the town without addressing community concerns about proximity to homes and impacts on property value, wetlands, and wildlife. What began as a few citizens in this rural township of 1,200 grew to over 150 members, and regular fund-raisers became community-wide events. In October 2011, I attended their fund-raising potluck and raffle, where more than a hundred people met in a local church. The food-laden tables offered a tempting array, from roast turkey and au gratin potatoes to fresh salads and lasagna, and a dessert table including homemade apple pies. The raffle basket I won was brimming with attractive items—pottery mugs and woven placemats, homemade pickled vegetables, and locally produced honey and maple syrup. Over four years, at potluck dinners, barbecues, holiday bazaars, and garage sales, CSOO sold T-shirts and raffle tickets, shared stories, celebrated unity—and raised over $200,000. The mission statement on the CSOO website reflects their common purpose: "We the concerned residents and non-resident taxpayers of the Town of Orangeville, in Wyoming County, New York State, are dedicated to preserving the rural beauty of the countryside along with the health and quality of life for all who live, work and play in our Town of Orangeville."[5]

Another successful citizens' group is Rockland Organized for Sustainability and a Safe Aquifer (ROSA), in Rockland County, New York. It represents residents in a county with a population of just over 300,000 and originally formed to protect undeveloped land over a major aquifer in a township of 126,500. Here is the ROSA mission statement: "ROSA's mission is to represent the interests of area residents in their demand for a safe, legal, sustainable, and well planned land use solution for the Patrick Farm property; a plan that will protect our shared water supply, the environment and the character of the community." ROSA's vision includes "looking to find reasonable and sustainable solutions that protect water supplies and community character." Drawing from a broad base of members, it has a board of directors, department managers, and work teams. The group has hired consultants and initiated legal action to protect the aquifer, funded by community donations. ROSA is a 501(c)(3) organization and accepts nonprofit donations via its website.[6]

The ROSA website also showcases testimonials from citizens in the section "We are ROSA."[7] Here is a sample:

- "I have chosen to support ROSA for . . . many reasons . . . including the effects such over-development will have on Water/Aquifer, Traffic, Environment/Ecosystem, Safety . . . Taxes, Home Values, Municipal Resources and Historic Preservation. But the main reason I am so incensed is that the town is not listening to its citizens." (Lisa L.)
- "I am very concerned about the loss/damage to wetlands as well as the larger environment. . . . The current wetlands and forest provide numerous ecosystem services to the local and regional area. . . . The developer's suggested housing and congestion are not compatible with the best use

of land and resources, and are counter to sustainable growth in the County." (Dorothy P.)

- "ROSA in representing us provides the necessary unity and organization, so we are not isolated voices. We become one strong voice in an effort to safeguard the Patrick Farm, protect our environment and maintain our way of life." (Sandra S.)

CSOO and ROSA are two examples of many coalitions across the country. From very small rural groups to sophisticated national groups with staff, coalitions play a vital role in bringing people together for effective action and providing funding to advance group efforts, whether it's a study to provide data or a retainer for litigation fees.

Meetings

Meetings help coalition or group members communicate, gather support, share or evaluate information, and brainstorm strategies. But no one wants to spend hours sitting in meetings that are too long and accomplish little. When you're trying to increase your base of volunteers, wasted time in meetings is an energy drain you can't afford. With some planning, meetings can be positive experiences that help volunteers stay energized, informed, and interested. To do this, here are a few tips to keep in mind:

1. State the purpose of the meeting for all who attend.
2. Choose a meeting place that has good seating, lighting, and accessibility. If people are physically comfortable, it's easier for them to relax and participate.
3. Designate someone to lead and facilitate the meeting. This person's job is to make sure that the meeting achieves its purpose, moves along, and stays on topic; that all present have a chance to participate; and that no individual monopolizes the conversation.
4. Prepare a written agenda. Even if it's fairly general, everyone starts the meeting literally on the same page. The agenda should include an ending time to make people more at ease about their time commitment.
5. Stick to the timing: start and finish on time. If those at the meeting want to continue past the ending time, you can extend it, with popular consent, at the end of the meeting. But give people who need to leave a chance to do so at the originally stated ending time.
6. Provide refreshments. These can be as simple as hot drinks in winter, cold drinks in summer, and snacks. Besides keeping people

awake and energized, refreshments are an outward show of hospitality. You are building community and showing people that you appreciate their time commitment.

7. Keep track of the business at hand. Designate someone to record what happens at the meeting. This written record updates those who miss a meeting and documents the group's actions.

These are general guidelines; specific meetings may require more or less formal organization, depending on purpose. People who are engaged in a well-planned meeting are more likely to participate and become involved in the group's actions.

If you suspect that a meeting may turn controversial because of a hot topic, or people who are angry and have an "ax to grind," refer to *Robert's Rules of Order* for specific advice. The more controversial the meeting, the greater the need for structure and tighter control by the person in charge. An assertive individual with facilitation skills is often a good choice for chairing a controversial meeting. The use of *Robert's Rules* can make the difference between a meeting's success or failure, showing you how to address any attempt to divert a group from its purpose. Have you ever watched someone ignore the purpose of the meeting, talk at length about irrelevant subjects, or become abusive of others? If left unmanaged, such actions can derail a meeting so that nothing is accomplished.

Here's a brief look at how *Robert's Rules* can help deal with incivility and prevent situations that undermine your purpose. Consult the full text for more details.

- *Breaches of order by members in a meeting.* This section describes how to call a disruptive member of the group to order and implement penalties for refusal to comply.[8]
- *Mass meetings* (for example, public hearings): "Any person at a mass meeting who, after being advised, persists in an obvious attempt to divert the meeting to a different purpose from that for which it was called, or who otherwise tries to disrupt the proceedings, becomes subject to . . . disciplinary procedures,"[9] which are described in subsequent pages.

Communications

Effective communication skills are critical to building communities and coalitions, presenting information so it's easily understood, and persuading people to support your cause. Communications include everything from conversations over coffee to social media. Some of the people you want to reach may be unavailable to attend meetings, so consider bringing information to them in a succinct, convenient form that meshes with their busy schedules. Be prepared to meet people where they are; make it easy for them to participate. Opportunities include the following:

- Maintain visibility in local media via letters to the editor, press releases, or local news.
- Build a website, blog, or Facebook page for your group; use social media appropriate to your issue. Post a "who we are" page on the group website where supporters share why they chose to become involved.
- Provide links to regulations, websites, and other pertinent sources— make it easy for people to become informed.
- Produce a newsletter, e-newsletter, fact sheets, or FAQs (frequently asked questions). Document your facts. Help residents interpret scientific data and complex reports.

How do you respond when someone asks why you're taking a particular position? With good research as a backup, you are less vulnerable to being caught in a breathless stutter when surprised by a spur-of-the-moment challenge that catches you without the facts for a good response (leading you to lament "I wish I'd said . . . at the time!").

Presenting Your Position

Know your audience! Whether you're addressing the public or local boards and elected officials, try to understand the position of the people you want to influence. Identify the fears behind individual or public reactions: What are people afraid will happen—or not happen—as a result of proposed actions? It's important to find out as much as you can about their concerns. With this knowledge, you can address those fears head-on. Awareness of your audience can make a big difference in how well they receive your message.

State your argument clearly to those who have the power to make decisions. How and when you present a case can be critical to success. When you are addressing a planning board on the deficiencies of a site-plan review and trying to show how a project would change a wetland's ability to function, it's not the ideal time to backtrack, define basic science terms, and explain how an ecosystem or wetland works. Most planning board members don't have a background in ecology—and because of time constraints, they appreciate simple, brief explanations. They don't have the time to investigate why a particular stream is important and how land-use activities will damage it. And if board members represent a town that is desperate for economic development and in a hurry to approve a project, you face an uphill battle when you explain why they should adopt a different action.

To anticipate this and avert crises, compile the pertinent data about water resources in your community before they are threatened. If this isn't possible, become involved as early as you can in the review process. Include in your strategy a time to educate local boards and officials by using a "bullet point" approach: short but pithy presentations, media clips, graphics, maps, fact sheets, and responses to FAQs.

Craft your responses in advance to those who oppose your position. Sometimes opposition is based on faulty information, and you can address this by providing the facts. Sometimes the real reasons are buried deeper, and you may have to dig to discover them. If the concern is about economic development, jobs, or local tax dollars and property values, match your response to those concerns. If you need dollar values for natural resources, find studies that evaluate ecosystem services. You can estimate the cost of replacing the lost natural resource or its benefits; for example, if you pollute a well, what would it cost to restore water quality to a drinkable standard or, alternatively, to replace it with another source of water?

At every possible opportunity, you need to address the true costs of not protecting natural resources. Be proactive and initiate the comparison. Use unbiased economic and jobs figures, and document your sources.

Perceptions

When you speak in public, be aware of how people you don't know will perceive you and your message. You may want to list some goals for yourself—for example, present an honest, forthright approach; give people a clear and informative message; convey your message from a position of strength. Avoid using labels like "pro-development" and "antidevelopment," which divide communities and create controversy. As someone who is concerned about protecting water resources, you can direct the conversation to specific issues and their solutions. If you oppose a project and are perceived as being "against" development in general, you create a less positive impression than you would if your approach was "for" protection of a stream's water quality and well-planned, sustainable development. Public perception of you or your group as fair and reasonable will benefit your credibility. Better to admit you don't know, and send someone to a good source for information, than to bluff through an inaccurate response to a question.

Another divisive label is political party. That division won't change overnight, but we can soften it by emphasizing unity of purpose. Your concern is protecting water, a common resource. This is fundamentally not a partisan issue, and anything you can do to stick to concerns everyone shares will help focus attention on the issues at hand. Whenever possible, redirect the discussion to water resource benefits for the community, threats to those benefits, and solutions. Remind people about interrelationships among water features, and the cause and effect of land-use activities that change them. This approach requires that your testimony is rooted in facts you can document.

We often hear negative statements about environmental protection from corporate and other interests that don't subscribe to the importance of the "common good." This approach can relegate concerned citizens (who question the effects of development activities) to a fringe group biased against growth and jobs—you're an extremist, environmentalist, tree-hugger, another special-interest group. This labeling creates a negative perception, or stereotype, fed by other similar statements. For example, the coal-mining industry dismisses concerns about the impacts of mountaintop removal mining as exaggerated and untrue. According to one industry executive, "What [environmentalists] are attempting to do is stir the emotions of people, when the facts are that the disturbance is limited, and the type of mining is controlled by the geology."[10] Such statements are meant to diminish the power and credibility of those who strive to protect water and other environmental assets. But these statements give us an opportunity to challenge negative perceptions—and create an alternative positive image of those who are concerned about natural resource protection.

If water is indeed a resource that we share and can't live without, then its protection concerns everyone. Protecting our environment—including water resources—is not limited to a special-interest group. Promote this awareness in your conversations, meetings, and public statements.

Perceptions affect our view of what constitutes an unbiased presentation. Shortly after accepting an invitation from a local group to give a presentation about hydrofracking and its impacts, I received another call, apologetic but stating that local officials compelled the group to contact someone else for the presentation because as an ecologist I was too "biased" in favor of environmental protection. Curious about my replacement, I attended that event. The presenter was a biologist who works for the oil and gas industry. His presentation was very well done but focused on a narrow definition of hydrofracking and omitted discussion of impacts. Afterward the group requested that I return and cover the "rest of the story."

Sometimes those who believe strongly in a cause can be carried away by passionate convictions—and risk feeding the negative stereotype. If your goal is to inspire support for your cause, creating a negative impression isn't a good strategy. Even if you're frustrated because you are personally affected by a project and no one seems to be listening, you need to stick to civil presentation and authoritative facts.

Using Different Types of Information

"Well-informed" involves understanding different types of information: regulatory, scientific, social, economic, and political. All these types of information contribute to your credibility and the effectiveness of your outreach efforts.

Understanding the Regulations

As noted earlier, government regulations do not guarantee environmental protection. You can maximize their effectiveness, however, by understanding how to use them. Become familiar with local, state, and federal regulations so that you are aware of all options for protecting resources, deadlines for review and comment, agency roles, and enforcement. A specific law may not be designed to address the specific water resource issue you're concerned about. You may need to consult several different laws to find the one that best matches your situation. For example, if groundwater pollution is an issue, you may need to look beyond the Clean Water Act—to the Safe Drinking Water Act, the Resource Conservation and Recovery Act, and the Superfund Act, depending on source and type of contamination.

Here is a quick overview of several federal laws that deal with some aspect of water resource protection. Refer to the EPA's website for a longer list and more details.[11]

- *National Environmental Policy Act.* Its intent is to ensure that federal government agencies consider the effects of their decisions on the environment prior to approving actions that may have significant impacts.
- *Clean Water Act.* Enacted to regulate discharges of pollutants into waters of the United States, this law establishes water-quality standards for surface waters. It includes provisions that address pollutant

discharges, stormwater discharges, nonpoint-source pollution, estuaries, oil spill prevention and control, and wetlands (Section 404).[12]

- *Endangered Species Act.* The intent of this law is to prevent listed endangered species from becoming extinct, listed threatened species from becoming endangered, and to identify critical habitat for these species.

- *Safe Drinking Water Act.* The act addresses quality of drinking water, including aboveground and belowground sources.

- *Resource Conservation and Recovery Act.* This controls the generation, transportation, treatment, storage, and disposal of hazardous wastes. It provides the framework for managing nonhazardous solid wastes.

- *Superfund Act.* The Comprehensive Environmental Response, Compensation, and Liability Act (CERCLA), as it is formally known, describes actions to address the release or threatened release of hazardous substances that may endanger public health or the environment. It establishes a fund for hazardous-site cleanup.

- *Toxic Substances Control Act (TOSCA).* This law provides testing requirements and restrictions for the production, importation, use, and disposal of specific chemicals (excluding drugs and pesticides).

In addition to these federal laws, states and municipalities have their own regulations and policies. Whenever possible, work with local boards and officials and enlist their support. Most board members are volunteers; as with all volunteers, respect their commitment of time, and help them access regulatory applications. When possible, build a working relationship with local officials and decision makers.

Even though regulatory jurisdiction supports your case, decision makers may be unable or unwilling to assist. There may be many reasons for this. But as a last resort you may need to obtain legal help to achieve your goals. Seek the advice of an attorney who is well versed in the regulations that apply to your cause. Other coalitions may be able to recommend legal resources in your area.

Information Sources

Consult the original source of facts. Word-of-mouth interpretation may change from person to person, like the whispered messages in the children's game "telephone." You may run a wild-goose chase based on what someone heard. When falsehoods muddy your issue, you must counter them immediately, with the authority of factual evidence. The smallest untrue statement can catch fire and spread, and the more it spreads, the harder it is to extinguish. Even if a statement seems obviously ridiculous to someone who is well-informed, those who don't know the facts may accept it. They need to hear from you! One person with sound evidence and communication skills can make a big difference.

Verify what you hear or read on a printed page, website, blog—unless you're sure it comes from a reliable source. When you present evidence to others, include its source—this is especially helpful in controversial situations, or when people are confused about the facts. If someone asks you a question and you don't know the answer, research a good response and put it in writing. If the question was asked in a public meeting, make sure the response reaches the people who most need to see it.

Attend local conferences and workshops; topics may include stormwater management, flooding, climate change, wetland protection, biodiversity, stream protection and restoration, buffers, aquifer protection, watershed management, better site design, and sustainable development. Local residents with no prior background can learn important details about wetlands, legal precedents, rare species, septic drainage-fields, industrial processes, and construction practices.

Using Scientific Information

Using scientific information to support your project or position requires a background in basic concepts, as well as quality control. Once you establish the facts, you still have to interpret what they mean in the context of your situation. And though we live in an age where more information is available to us than ever before, there is little or no quality control.

When you understand the science about natural water systems, you are less likely to be fooled by assurances from those who may not be using the best available facts. Think in terms of watersheds, ecosystems, and movement of water aboveground and belowground. Become familiar with the system you're trying to protect, whether it's a wetland, stream, watershed, or aquifer. Be able to describe its services and benefits.

Use natural-resources science to build a convincing case. Collect evidence to prove why land-use impacts are "significant" and deserve greater attention, or why proposed mitigation won't work. Bring up specific examples of the effects of cumulative impacts.

The presence of a listed threatened or endangered species can help your cause, but remember these species are part of a larger ecosystem that provides additional benefits for human communities. Focus on those benefits if at all possible. It may be unwise to base your entire case on impacts to one endangered species, but sometimes the presence of this species can help you get the attention and assistance needed to make your point. Enlist the aid of an expert to verify photos or other documentation of species presence.

Document your observations with photos, maps, and other visual aids. If you're taking photos, make sure you provide either a GPS reading, a video recording, or a map indicating the photo location with time and date.

When you understand scientific facts, you can craft an effective response to counter misinformation. You can be tenacious when challenged if you have documented evidence to support your claims and cite credible sources.

Using Economic, Social, and Political Information

Sometimes an elected official promises to base a decision about a controversial project on science alone—objective facts, not clouded by opinion or emotion. This sounds like a good idea, and it usually holds up as long as the science-based decision is easy. But how well do you think that approach prevails when the facts are telling you that a particular job-generating, tax-income-providing development project will contaminate your water?

Science (even when it comes to public health issues) rarely wins over promises of economic gain—whether or not those promises are true. There are occasional exceptions; for example, the recent decision to ban hydrofracking in New York State after the release of a comprehensive public health review. Significant public input also played a role in that decision.

Information about our health and the natural resources we value is one piece of a mosaic that also includes people with competing interests and different values as well as economic realities such as the need for housing, jobs, and reliable sources of energy. While scientific facts are the foundation for environmental protection, a good strategy also requires a solid understanding of political, economic, and social realities.

Controversial projects can divide communities; media headlines target disputes and issues that divide people into opposing camps. But what are people really arguing about? The issues behind these stories may include industry relationships with regulatory authorities, political agreements, insufficient tax dollars, or lack of jobs. People with competing interests have different ideas of what is at stake when land-use decisions are made; take them into consideration as you develop your strategy.

Become familiar with land use in your municipality. Locate waste sites, pipelines, wastewater treatment facilities, lands zoned for intense use, mining, or industrial sites. Understand the costs of the additional municipal services (schools, fire protection, police, road maintenance, water treatment) required by an influx of new residents or the construction of an industrial complex.

People like to be reassured. It's hard to believe that a government entity or a corporation would knowingly permit actions that could harm residents or a community. But it is naive to expect government regulations or corporate goodwill to take care of the problem and give us all the information we need for good local decisions. Document the consequences of development activities, including who is responsible for bearing the cost of cleaning up contaminant spills or replacing polluted water supplies. These facts may not be popular amid anticipation of more tax money or jobs, but they do merit local evaluation. As we have seen in previous chapters, ignoring the causes of our water problems doesn't solve them or make them go away; but it does allow them to get worse over time.

Proactive Options

You don't have to wait for a controversy to develop before taking action to protect your water resources. Developing a strategy for local resource protection can breathe new life into reports and plans gathering dust on a shelf. By planning to protect sensitive resources before they are threatened, you can improve existing project review processes, reduce controversy, and better protect the environment.

A proactive strategic plan may include guidelines for developers. For example, the planning department in Larimer County, Colorado, initiated a project to identify and mitigate negative impacts of development. As in many other places, local developers planned projects without knowing in advance what the sensitive or significant local natural resources were, where they were located, and how the county wanted to protect them. So they would proceed with project design and engineering drawings, only to be told after the fact to make major changes to accommodate local concerns about important resources. This incurred additional expense and created conflict between the county and developers, who blamed local government for onerous environmental protection requirements. A set of "Developers' Guidelines" was drafted as county policy, informing developers about avoiding impacts to important natural resources before they prepared expensive site plans. This proactive approach appealed to county planners, environmental protection advocates, and developers; though the developers grumbled at first, some of them admitted it was helpful to know the "constraints" to development ahead of time.

You can incorporate a variety of proactive projects, studies, and outreach actions into a strategy for environmental protection. Be creative!

Here are some ideas to get started:

1. Compile and display an atlas, or inventory with maps of local watersheds, wetlands and streams, and other water resources. Describe how they work, how they're interconnected, and how land-use activities threaten them. Make the atlas visually attractive (use photos) and easy to use.

2. Develop a local law, zoning code, overlay, or ordinance that protects water resource systems and their buffers.

3. Design and produce a state-of-the-resources "scorecard" for water resources of special significance like CBF's State of the Bay. This can monitor changes over time, attract the attention of both the public and elected officials, and provide a basis for protection actions.

4. Engage the public by sponsoring a photo contest—for example, best scenic water feature, best water-recreation photo, favorite stream, or interesting aquatic creatures. Increase the visibility of the photos by displaying them in a public place or creating a photo book. Generate news articles to publicize the contest and its results, and come up with some creative prizes.

5. Consider making a video to illustrate and explain your topic of concern (or raise funds to have one professionally produced)—it can be short, but it should be as professional as possible. Pay attention to quality, details, and approach. A video has the potential to make or break your ability to capture the public's attention.

6. Collect articles, stories, and other information for a newsletter, e-newsletter, website, series of fact sheets, or Facebook page about the water resources in your community and why they are important. Here are a few ideas of what you could include:
   a. Information on featured ecosystems or watersheds and the services they provide
   b. Highlights of special local water features, habitats, species, or scenes
   c. Tips for linking project activities with specific natural resources that may be affected

d. Facts about specific water contaminants, cumulative impacts, or other water-resource concerns

e. Articles showcasing a local wetland or stream restoration effort

7. Organize nature walks or other activities that bring people into nature and show them what you would like to protect. Include local volunteer groups and elected officials.

8. Support organizations that are working to improve the environment. Help to elevate education about natural resources (for all ages) to a high priority. Many conservation organizations provide this information; they can all use funding and volunteers.

9. Develop guidelines or promote policy for local environmental reviews. Consider the following examples:

a. Developers' guidelines

b. Habitat-assessment guidelines

c. Local requirements for low-impact development and sustainability

d. Local criteria for identifying significant impacts

e. Methods for evaluating "true costs" in project reviews

f. Criteria for evaluating mitigation success

10. Encourage conservation easements. Support local land trusts. Sometimes lack of funding and the need for long-term monitoring discourage participation in easement programs. A land trust or agency entrusted with administering the easement requires staff. The up-front payment required to set up a conservation easement can be daunting to landowners, so local actions to help cover these costs can encourage more people to participate.

11. Promote buffer protection. Plant trees in buffers adjacent to streams, rivers, lakes, and wetlands. Create a local buffer planting project; guidelines are available from state and federal forestry agencies.[13]

12. Promote innovative techniques for managing stormwater and wastewater. For example, the Omega Institute in Rhinebeck, New York, has developed the Center for Sustainable Living, an environmental education center and natural water-reclamation facility in the first green building in the United States to receive LEED Platinum and Living Building Challenge certification. At the center, the "Eco Machine" treats wastewater without using chemicals, and solar and geothermal systems provide energy, heating, and cooling for the building. All the water from Omega's campus, including water used in toilets, showers, and sinks, flows to the Eco Machine, where it is purified by microscopic algae, fungi, bacteria, plants, and

snails. The purified water returns to the aquifer via large dispersal fields under the parking lot.[14]

The Omega Institute sets a high standard, worth considering as a good example of what is possible and moving us beyond the limitations of "this is the way we've always done it."

Staying Power

Value of Organization

Keeping track of reports, e-mails, and contacts over the course of a project can be daunting. If your project continues over months or years, "documentation" can accumulate to the point where important details may be lost. To avoid losing files, from the outset organize a place for all records of phone conversations, e-mails, contact information, websites, experts, legal documents, and reports.

Two Hudson Valley neighbors who own portions of the same large wetland practice "organization in action." When a developer approached the town with plans for a high-density residential project on an adjacent undeveloped parcel, the neighbors organized to protect the wetland and its buffers. They set up space in a spare room off the garage, dubbed the "War Room," with bookshelves, computer, whiteboard, and a large table covered with reports and letters, all tagged, labeled, and organized. Contact information for agency representatives, public officials, other neighbors, and various experts was immediately accessible. In this room residents held meetings, called people, gathered and organized information—and occasionally raided the small refrigerator to share refreshments and a few jokes. One landowner constructed a small lean-to at the edge of the wetland, well situated for observing birds and turtles. Barbecues were hosted at the "Turtle Hut," bringing supporters, and sometimes local officials, together for a firsthand look at the wetland. The wetland protection team used it as a reminder of the resource they were working hard to protect.

All of this organization was necessary for sustaining the wetland protection effort, because the project review continued for several years through controversies, revisions, and eventually legal action. It cost tens of thousands of dollars, in this case fronted primarily by the two neighbors who hosted the War Room and the Turtle Hut. The developer alternately dropped, redesigned, reintroduced, and redrew the project; the current version has a smaller footprint, farther from the wetland.

Now a mining company is stripping trees from a nearby parcel and filling wetlands that feed into the system described above. The need to organize for action continues. To meet the new challenge, residents already have background information about the area's natural resources and environmental protection issues.

Never Quit

How long does it take to achieve your goals of protection? That may depend on how you define "protection." For over twenty years a landowner in a rural Catskill community has worked to protect a several-hundred-acre wetland complex in a wide valley. She learned to identify wetland plants and soils, understand construction practices, and evaluate septic system sites. She also encouraged her neighbors to protect wetlands and small streams on their properties. Her home is a library of maps, documents, reports, and photos. Reflecting on the years she has spent on this project, she sighs and says that she doesn't seem to have made much progress, because inappropriately sited development continues to threaten the wetland. But her actions, energy, and sheer persistence have raised local awareness; her neighbors have become educated about the need to protect local water resources; and now she's working to encourage local conservation easements. Her influence is long term; state and federal agencies are aware of the local wetland issues, and the state has amended some of its wetland mapping to include more of the wetland area, though comprehensive protection has not been forthcoming. Though residents experience recurring and often severe flooding, local boards continue to allow the filling of unmapped wetlands and small streams.

What does perseverance mean to you and your local efforts? Imagine: Five years have gone by and you have invested time and money to protect natural resources you value. Your case ended in a legal battle that is still in court. A friend tells you she tried to oppose the same type of development project in a nearby municipality, but her attempt failed and she lost the case—so, she says, any continued attempts on your part to prevail in a similar effort are likely to be futile. People who face a large corporation might say "We can't fight this; the outcome is inevitable. This corporation has so much money it can control the whole process. We know there's a problem, and it's affecting our health, but there's nothing we can do to stop it."

But shared adversity gives us opportunities to take heart from each other's efforts and successes. For example, the hydrofracking controversy has given rise to a growing number of citizens' groups and coalitions in states across the country. In the spirit of "never give up," several groups in New York adapted the statement of purpose from the *No Frack Almanac* and displayed it on their websites:

> Many of us who live here think these industries are dangerous and will destroy our beloved countryside. We are trying to stop these projects, but it is an unequal struggle. The gas companies have spent literally hundreds of millions of dollars on lobbying and advertising to promote their plans. No one in the opposition has that kind of money, but we do have something the fracking companies don't have, and that's millions of people who love our area just the way it is. In *The No Frack Almanac* we will tell our side of the story, the side that's not advertised in 60-second commercials on TV. We hope we can tell you enough to make you want to find out more and, ultimately, to help keep our area, or any area, from being destroyed.[15]

This statement and the public involvement it reflects are especially meaningful in light of New York's eventual ban on hydrofracking. But the need for continued public involvement is far from over. In New York, organized citizens' groups will be needed to address natural gas infrastructure, transportation, and other energy development that continues to threaten the environment. In the more than twenty states where fracking is currently in practice and several additional states where it has been proposed, citizens can benefit from the work of New York's coalitions.

As population increases, so does competition for water. Threats to natural water systems will continue to grow, and the need for grassroots efforts to protect these resources will become ever more urgent. Knowledgeable advocates for water resources protection will always be needed, because we will never be able to take that protection for granted. We can dispel the notion that concern for the environment is a "special interest" or political position. The key is to focus on the importance of protecting water resources and never let up. Keep asking questions.

Working to protect the environment is a marathon. Plan on it. Just like conditioning and training to run twenty-six miles, you need to prepare with data, coalitions, communications, and strategy—and understand how to reenergize yourself as you go. If you build community as part of your effort, enjoy the people who share your concerns. Passion for a good cause helps to keep the energy level up; it's the fuel for the fire that results in action. Many are convinced they can't possibly make a difference. But if not us, then who will? On their own, governments, corporations, or elected officials are not able to choose the path to a healthier future for all of us.

Water resource protection—or any other type of environmental protection—requires more than plans, reports, and regulations. It requires you: your energy, passion, and commitment to your community and its future.

Conclusion
Stars in the Water

Unless someone like you cares a whole awful lot,
Nothing is going to get better. It's not.
DR. SEUSS, *The Lorax*

Flying from Colorado to New York on a clear spring morning gives me a great opportunity to look at land and water below and reflect on their condition. Today it's easy to see—from the forested foothills of Colorado's Rockies, into the eastern plains—a patchwork of irrigated fields, circle sprinklers, and grid-pattern roads. Meandering rivers snake through forest patches, fields, and residential developments. From this aerial vantage point I can see the network of rivers and the patterns of land development. It's a big country; I can easily be lulled into thinking the landscape below me is so big that human activities can't possibly damage it. As we gradually drop altitude approaching Newark, the patterns below me expand, and I can see more details. Much of the land here is developed, with more roads, parking lots, high-density housing. Remnants of the original network of streams and rivers lack even a fringe of riparian vegetation. They may have shaped early development, but now intensive land use has changed them dramatically. As the plane touches down in Newark, my perspective from the air shifts to what is right in front of me: this building, that road, a group of forlorn trees, and trash at the edge of a muddy pond.

Shifts in Thinking

Billions of people live on this planet. Our activities and lifestyles inevitably change natural resources, producing some effects that are irreversible and others that can be mitigated. We are part of both the natural system we try to protect and the forces that change it. We are challenged to support sustainable communities, energy sources, and lifestyles. Meeting this challenge is a matter of information, priorities, and will, but it also requires a shift in our thinking about how we value the environment and our place in it—a major change that takes time.

But time is not on our side as we face the effects of climate change, including warmer temperatures, changing rainfall patterns, more frequent severe storms, more flooding, and increased ocean acidity. We see ecosystem changes in timing of leaf-out, flowering, and fruiting; altered stream-flow and hydroperiod; drought; shifting animal migration patterns and breeding cycles; varying food availability; and altered nutrient cycling. We can predict physical changes *for each degree of climate warming*:[1]

- Precipitation increases or decreases by 5–10 percent in many regions
- Average September extent of Arctic sea ice decreases by about 25 percent
- Rain during heaviest precipitation events increases by 3–10 percent
- Stream flow decreases by 5–10 percent in some river basins (with a decrease in watershed runoff into surface waters)

As we grapple with the reality of climate change and a shift in how we think about energy sources, we're also challenged to address the need for more effective environmental protection. Cumulative impacts are piling up to wreak serious havoc on our land and water.

One way to measure this cumulative effect is the "ecological footprint." The footprint compares the earth's ability to produce natural resources and absorb wastes with our demand for natural resources and our production of wastes. While the earth provides the renewable resources we use and can absorb many of our waste products (including carbon emissions), natural resources are limited, and so is the amount of waste that can be absorbed. All human activities and impacts that make up the ecological footprint have a cumulative effect. World Wildlife Fund research indicates that by the 1980s our global ecological footprint had reached the earth's capacity to provide resources and absorb wastes, and by 2003 we had exceeded that capacity by 25 percent.[2] Americans have perhaps the largest per capita ecological footprint in the world; scientists have documented extensive ecosystem degradation across the country.[3] According to the UN Millennium Assessment, 60 percent of global ecosystem services are being degraded or used at levels that can't be sustained—including freshwater supply, fisheries, air and water purification, and the regulation of natural hazards and pests.[4]

So how do we elevate the protection of the environment to a high priority? How can we change our collective thinking to view environmental conservation in terms of protecting natural systems from cumulative impacts? While the earth has a limited capacity to deal with degradation, our regulations and land-use decisions favor a cost-benefit scenario that places a priority on economic development over natural resource protection. The consequences of this are clear from an aerial view, as well as from a reconnaissance on the ground. But a key to the solution—that dramatic shift in thinking—doesn't seem to be in the works, at least not now. It seems as remote as the landscape viewed from thirty thousand feet.

The need for a change in the way we value and protect natural resources is not new. Committed conservationists have been working on it for a long time. For example, Marjory Stoneman Douglas championed protection of Florida's Everglades against efforts to drain and develop them. Her book *The Everglades: River of Grass* (1947) helped change the perception of the Everglades from worthless swamp to valuable natural resource. In 1993, when she was 103 and still active, she was awarded the Presidential Medal of Freedom. The citation reads, "Her crusade to preserve and restore the Everglades has enhanced our Nation's respect for our precious environment."[5]

Even as we celebrate Douglas's fine work, the battle to protect Florida's Everglades is probably not over. Challenges will continue to arise. But it's important to remember her story, and the stories of others who work hard to protect the natural resources they value. Each success is a contribution to the shift in our collective thinking.

I began this book remembering how optimistic I felt about environmental protection when I was in college. Pete Seeger had recently launched the sloop *Clearwater*, sailing the Hudson River in a crusade for clean water. Today his work continues through the Clearwater organization, and the sloop still sails. Through his gift of song, Seeger brought people together, focused on action, and changed our view of the river. In 1994 he said, "The key to the future of the world is finding the optimistic stories and letting them be known."[6]

Success stories alone may not shift the balance toward environmental protection, but they do have the power to energize and inspire us, both as individuals and in communities.

Many or One?

Coalitions and communities that amass enough power to make a difference are composed of many "ones." Our personal energy comes from recognizing that power of the individual. Ironically, when many "ones" come together to form a coalition, they become a "super-one" in terms of what they are able to accomplish. But what can you do as one individual? We looked at some options in the last chapter. Ask questions. Become informed. Support environmental protection organizations. Get involved in local efforts to protect natural resources.

It takes one person to ask the right question, bring up a topic everyone else has ignored, talk to a neighbor, do some research, write a letter, create a piece of artwork, start a coalition, or remind us of a success story. We see the results one small victory at a time—a neighbor who opposed your cause until you spent time talking about it one on one; a photo contest to raise local awareness; the first person to throw a pebble into the calm pool of complacency and make a few tiny waves.

The "power of one" has many faces. One of them is yours.

Preserving Your Land

We can limit some of the negative effects of land-use activities on water by protecting land, one parcel at a time.

The pond behind my childhood home in the Hudson Valley is still there. We had to sell the property about ten years ago—a tough proposition for me, despite the logic of it. After my father's death, my mother needed to move. I insisted on setting it up as a conservation easement with the local land trust before putting it on the market—though I had to fight to convince the real estate agent, who was certain we would never get the asking price, and even my mother, who listened to the advice of the agent. My passion for the setting that defined my childhood fueled my stubbornness. With bittersweet relief I received an offer for the property that was exactly the asking price, and the folks who made the offer were thrilled with the easement and the property's natural features. Ironically, today I understand the ecosystem value of the place; as a child I simply enjoyed it.

Do you love a place outdoors, a place with water? It could be a scene from your childhood, the place where you live now, a backyard, garden, a nearby undeveloped park, vacation spot—or even a wild place you've never seen in person. What can you do to protect your connection with these places? The experience of setting up a conservation easement reassured me that the property continues to be valued—and the natural world that nurtured my own growth is allowed to flourish. At first, I thought that was all I could do. Today I think I can do more. Special places—other ponds, streams, and woodlands—have the potential to make a difference in individual lives. These places need to be the next conservation easements, refuges for future generations of kids who love to poke around in the mud and revel in the intoxicating variety of life. Advocating for those places can be an opportunity to give something back. My parents gave me a great childhood in a special place. I hope its impression never leaves me.

The conservation easement movement is growing. The National Conservation Easement Data Base (NCED) website reports the number of conservation easements, as of September 2013, at 101,203, for a total of 19,805,669 acres.[7] A public-private partnership, the NCED compiles records from land trusts and public agencies and shares easement information by state and county. Maps on the website chart the "conservation footprint" of easements across the country.

Power of One

In his essay "The Star Thrower," the naturalist Loren Eiseley describes a beach strewn with starfish stranded by a storm. Shell seekers up at dawn comb the surf's edge for other stranded creatures—plucking them from the water, filling their buckets, and even tossing live creatures into boiling pots to yield shells free of their unlucky inhabitants. Apart from the fray, on a deserted stretch of beach, a lone beachcomber is moved to return the live starfish to the surf, so he flings them oceanward, one by one. But there are too many. He knows that even of those he tries to rescue, only a few, or maybe only one, will survive its ordeal on the sand. So he has to content himself with that consolation: even if he saves just one, his effort will have been worthwhile. His personal choice makes a difference to each starfish he saves and reflects his view of the natural world and his place in it. This has nothing to do with saving a population or an endangered species; it has everything to do with how we look at ourselves in relationship with natural systems—and with the individual living plants and animals that populate those systems.

Multiplier Effect

Sometimes our individual actions can save more than one. Watery ecosystems across the country give us multiple chances to make a difference. In the land of vernal pools, the first rains of early spring bring spotted salamanders out of their underground burrows, and they begin the annual migration to the small wetland pools where they breed. Their striking yellow-spotted black bodies trek through the leaf litter, sometimes through a skiff of snow. For a creature six to nine inches long, it's an extensive march—four hundred feet or more through the forest to the pool. Some individuals have been migrating to and from these pools for twenty years. But problems arise when this march takes them across roads that separate forest from wetland. At well-traveled crossings, on a migration "big night," large groups made up of several species of salamanders and frogs may cross in the same place at the same time—ten, twenty, a hundred individuals determined to reach that breeding pool. At big crossings, death on the road can take a large toll. In the Hudson Valley, New York State's Woodland Pool Project brings volunteers out in the late-March drizzle, flashlights trained on salamanders that must be carried safely across the road, one at a time. Volunteers record size, location, and weather conditions. This information can be used to identify wetlands that are important breeding sites, and to plan future crossing assistance.

But there's more to the story of these salamanders and the wetlands where they breed; they illustrate "connecting the drops"—the water and land connections discussed throughout this book.

The salamanders are indicators of a healthy ecosystem. In the aquatic stage, larval salamanders feed voraciously on insects in the wetland, including mosquito larvae. The presence of the larval salamanders influences the biology of the pool; eventually they transform into the adult stage, climbing out of the water and dispersing into the surrounding woods. There they hide under logs, leaves, and soil—feasting on worms, slugs, and insects until next year's migration. In turn, salamanders are prey for some of the mammals, snakes, and birds of the woodlands. The salamanders connect the aquatic wetland habitat with the terrestrial, wooded habitat. The wetlands where salamanders lay their eggs are part of a watershed; wetland-contributing drainage areas include the woodlands where the adult salamanders live. Salamander habitat requires trees that maintain ground moisture and provide leaf litter that attracts salamander food. The trees also provide watershed functions like water-quality improvement, erosion control, and groundwater recharge. Besides their habitat function, vernal pools are part of the wetland mosaic across the watershed that stores water; collectively, they decrease the effects of flooding and release floodwaters slowly into the ground. Some are connected directly with groundwater. If their water becomes contaminated, they may not be able to support salamander eggs and larvae. When we fill in vernal pools throughout a watershed, we diminish its capacity to store water.

Protection of breeding pools and their surrounding woodland is as important as rescuing the individual migrants from death on the road. It takes education and a year-round effort to protect this system—keeping the pools intact without the adjacent woodland won't sustain the salamanders. Filling in the pools affects not only the salamanders but also the watershed.

Reaching to the Future

In this book we've looked at protecting natural water systems. Watersheds collect and move water across the land. Ecosystems within them purify and store water; sustain plants and animals; power nutrient cycles and transform the sun's energy into leaf, fin, and food. And as we identified the effects of our activities—how we move earth, change vegetation, and pollute the water—we also considered at what point an impact is "too much" for an ecosystem to process or when it interferes with watershed services to the extent that it should be prohibited. But in scrutinizing some of the obstacles to protection, and figuring out how to maximize our chances of successfully protecting our water, we've neglected an important detail that relates back to the vernal pool salamander migrations. Some of the volunteers helping to move salamanders out of the road are kids. Fascinated by the march of the amphibians, they eagerly participate in the night adventure. Through this hands-on experience, they learn about the larger system that supports it.

Imagine the power of reaching out to children and recognizing their capacity to protect water resources into the future. The notion that our children are our future is not new; a Native American proverb advises: "Treat the earth well: it was not given to you by your parents, it was loaned to you by your children."[8]

Teach Your Children

Childhood experiences affect our attitudes about natural resources throughout our lives. Memories of these experiences can resurface many years later. When a small town decided to amend its zoning code to provide better protection for streams and wetlands, residents who attended the informational town meeting were invited to share what they valued about water. The positive conversation helped dispel some of the controversy about the proposed changes by putting a personal face on natural resources and identifying some of the values residents shared. One elderly woman, who grew up in that town, was vehemently opposed to any zoning change and didn't want to talk about values. But eventually her neighbors helped her remember a scene from her childhood. She recalled playing in a small stream on her family's farm; she was especially fond of heading to the water with her pet calf to cool off during the heat of summer, standing in that special shady spot, with the water swirling around her legs. Her memory affected her perspective in the present. Once she remembered that stream's personal value to her, she was willing to listen to the plans for protecting it.

What will we give our children to remember about their connection with nature? Will they have positive experiences to draw on later in their lives?

One of the biggest obstacles to protecting water resources into the future will be a shortage of adults who care that these resources are protected.

In the short documentary *Project Wild Thing*, the filmmaker David Bond tells the story of his efforts to get his son and daughter outside and how he becomes a "marketing director for nature, working with branding and outdoor experts to develop a campaign." Why have children, whether they live in cities or the countryside, become disconnected from nature and the outdoors? The reasons are complex. Bond tells us *Project Wild Thing* "isn't some misty-eyed nostalgia for the past. We need to make more space for wild time in children's daily routine, freeing this generation of kids to have the sort of experiences that many of us took for granted."[9]

The increasing divide between young people and the natural world has environmental, physical, social, psychological, and spiritual implications. Research reveals the "necessity of contact with nature for healthy child and adult development."[10] How children connect with nature, and pass this on to their children, will determine how well the environment is protected into the future. Getting kids outside and helping them connect with the natural world isn't an easy task. And there are always unanticipated obstacles.

Picture this. A group of elementary school students on a field trip board a green bus labeled "Meet the Trees Foundation." The trip leader is dressed like a forest ranger. As the bus moves along, he shows them pictures of tree leaves. The presentation is not dynamic. The students are obviously bored, silent, restless; some are even nodding off . . . when voila! The leader whips off his ranger shirt to reveal a Toys-R-Us T-shirt and gleefully announces, "We're not going to the forest today—we're going to Toys-R-Us!" The kids, immediately transformed, not only wake up; they scream and cheer with delight. The bus rolls into the parking lot and confetti flies, as kids swarm into the store. While they are ecstatically grabbing and playing with an abundance of colorful plastic toys, the voice in the background invites us to "make all their wishes come true!"

This is an actual ad run by Toys-R-Us in 2013. I'm not picking on Toys-R-Us; I've been a customer myself. But this issue is far more widespread than one retailer. It reflects our widespread devaluing (and underfunding) of science education. Despite the stellar efforts of environmental educators, often working with very limited resources, we are not preparing enough children for a role in protecting the natural world that sustains them. We are not even passing along vital information about the fact that it does sustain them.

The Last Great Place

Water can ease the challenge of reaching kids. Children have a natural affinity for getting wet and muddy, and this continues as they get older, from splashing around to swimming, water-skiing, fishing, and paddling a canoe. That is, when they get the chance. Most of us now live in urban or suburban settings with a bike path or ball fields and limited access to natural places. With busy lives, we face a challenge in just getting kids to the water—not a swimming pool, but a natural pond, lake, wetland, or stream (one that's clean enough to play in safely!). Whenever you can find the time, the place, and the wherewithal to pry some young person away from electronics and get him or her outside, even for a little while—this is a worthy effort.

I brought my two granddaughters, ages eight and ten, who live in Colorado, to the Verkeederkill Falls on New York's Shawangunk Ridge. It is, as the Nature Conservancy sign proclaims, "One of the Earth's Last Great Places." After a rigorous, meandering hike to the top of the falls—with its impressive 180-foot drop—the girls were especially interested in that sign: "What does that mean, 'Last Great Place'? Why is it called that?" My reply was short; mostly I let the ridge work its magic. At that point, my job was to guide them to the experience. The swimming hole, fed by the stream that later plunged over the edge as waterfall, was awesome! Like hungry bears we stuffed ourselves with ripe blueberries along the stream's edge.

I didn't sell the view. I didn't push the swimming hole. All I did was get them there and leave the door open for them to discover a personal connection. I held back and let them move ahead on their own to explore upstream, where the channel narrowed to a leafy tunnel framed by highbush blueberry, and shallow water slipped over smooth rock slabs. For half an hour, at the top of the mountain, my granddaughters had their own adventure.

The missing piece in the "environmental protection = information" equation is a personal connection to a place or experience that makes an impression to last and to motivate. You can't get this connection just from reading about it or watching a video. You have to be there, to feel the exhilaration of a splash of cold, clear water after a hot, sweaty hike, the smell of the ferns, the soft coolness of green moss where you leave bare footprints at the stream's edge. In *Last Child in the Woods*, Richard Louv writes: "While knowledge about nature is vital, passion is the long-distance fuel for the struggle to save what is left of our natural heritage. . . . Passion does not arrive on videotape or on a CD; passion is personal. Passion is lifted from the earth itself by the muddy hands of the young; it travels along grass-stained sleeves to the heart."[11]

I have brought all five of my grandkids to this stream's cool, clear waters and watched them hop the boulders, play with toy boats, and fall in. As adults, will they care enough to preserve and protect nature? Will they know that the future of clear clean water is in their hands? Maybe not yet; but eventually, I hope. After all my emphasis on the importance of facts and credibility throughout this book, I must also stress that passion inspired by personal experience is what makes the natural world come alive for people. Take any opportunity, however small, to show that world to a child. It doesn't have to be a rare species or a "last great place"—just a window into a natural scene (best if it contains water!). Why do you want to protect it? Share that. The possibilities are endless.

The Story of Water

Every positive action to protect water resources is like tossing a pebble into a still pool of water: the ripples spread and grow. Maybe you capture someone else's interest in a good story about what you did or inspire a neighbor to take action. You can have an influence that extends beyond your backyard.

Many of the stories in this book are about real people who have decided to take action to improve their environment. They represent a fraction of the stories about individuals and groups across the country, engaged in human dramas to protect the natural settings they love. Collectively our storytelling of how we live on Earth involves everyone. We all need clean air, clean water, food, and shelter, and as we strive for these things, we create habitat and communities for ourselves. People have learned to live in deserts, tundra, rain forests, on mountains, in valleys, and on plains. Each place we live, we settle among the plants and animals that evolved in that place and make it unique. We take what we need from natural systems and eventually change the original community. But no matter how much we change, or where we live, we still need the basics: air, water, and food.

Water is intrinsic to our story. But the future supply, in many places, may be sharply diminished from what it is now. A study from the Cooperative Institute for Research in Environmental Sciences (CIRES) at the University of Colorado–Boulder evaluated the supply and demand for freshwater resources in 2,103 watersheds in the continental United States. The lead author, Kristen Averyt, reports, "By midcentury, we expect to see less reliable surface water supplies in several regions of the United States. . . . This is likely to create growing challenges for agriculture, electrical suppliers and municipalities, as there may be more demand for water and less to go around."[12] Looking at water supply from a regional perspective, CIRES found the following:

- Demands for water are greater than the natural supplies in 193 of the watersheds evaluated.
- In most parts of the country, agriculture requires more water than other activities.
- In some areas (such as Southern California), water stress comes from the water needs of large cities.
- In other locations, electric power plants present the largest demand on water (used for cooling).
- The western states are especially vulnerable to water stress—mainly because the difference between supply and demand is small, and users have relied on imported or stored water from other watersheds to supplement the natural supply and meet water needs.

Conditions in the Colorado River basin illustrate these trends. A 2014 satellite study by NASA and the University of California revealed that more than 75 percent of water lost in the Colorado River basin since 2004 came from groundwater. The extent of this loss may pose a greater threat to the water supply of the western United States than previously thought. "We don't know exactly how much groundwater we have left, so we don't know when we're going to run out," said Stephanie Castle, a water resources specialist at the University of California–Irvine and the study's lead author. "This is a lot of water to lose. We thought that the picture could be pretty bad, but this was shocking."[13] The US Bureau of Reclamation manages surface water in the basin's rivers and lakes and documents water losses. However, it is up to individual states to regulate pumping from underground aquifers, which is often not well documented.

Our relationship with water quickly turns to drama when we compete for it, sell or buy it, lose it, change it, steal it, and debate how (or whether) we will protect it. This drama displays the best and the worst of human nature, and everything in between. On the positive side, we have made some dramatic advancements. For example, in 1896, after construction of Seattle's first public water supply system, the city's leaders developed a long-term plan to own the entire one-hundred-thousand-acre Cedar River Watershed. This action would permanently protect Seattle's source of drinking water. After one hundred years, the city achieved this goal. "Seattle made a cost-effective investment in clean source waters that will never be threatened by roads, sewers, or urban runoff. It is an investment that will pay many times over through reduced treatment costs and a safe supply of water for generations to come."[14]

Why would we be timid about insisting that our water ought to be protected? Admittedly, it's a tough job. Water is constantly on the move—flowing everywhere, above and below the ground, too much in one place and not enough in another, distributing contaminants it picks up along the way. But this is the challenge, not an excuse to give up. Each of us has a fundamental human right to a clean, adequate supply of water. We will all foot the cleanup bill for poor land-use decisions. Worse, so will our children. Where the story of water goes is in part up to us, and how persistent we are in "connecting the drops."

Results may take a long time to become visible, and in the meantime you might wonder, "Why am I doing this?" But it helps to remember where you started, what caught your attention in the first place. My experiences with water—mucking around in countless swamps, fishing with my husband and sons, canoeing lakes and rivers—give me a wealth of positive water experiences to draw from.

I admit that I'm motivated by a visceral concern for the watery places I love. I am presenting this book to you as a challenge in the style of "swinging the pendulum back," as my father presented it to me. It's also an invitation to get out in nature and enjoy the benefits of water—whether it's a paddle in the river or just a glass of pure, cold water on a hot day.

Stars in the Water

When my son Matt was three, I took him out in the canoe on the Poudre River, which ran behind our house in Laporte, Colorado. Shadows of trout slipped out of sight as we glided along, sunlight on water. Matt pointed a small finger at the water in amazement, announcing, "'Tars, Mom! 'Tars on the water!" And so there were—flickering, starry pinpoints of light dancing on the achingly clear water and projecting themselves as watery stars on the stones that stood out in clear relief on the bottom. Light. Stars. Clear water.

As we look into the water, our vision is too often muddied by a confusing mishmash of daily events and the realities of our times. Pollution, drought, floods, energy needs, climate change, drinking water. What can we do to clear the vision?

When I look at the size of the problems confronting us, I'm almost overwhelmed before I begin to envision a way forward. What keeps me going? Glimpses of clear water, like the one I just described. I wish you many such glimpses. Let's aim for seeing stars of light shimmering in all of our waters.

Show a young person the stars in the water.

appendix
Stream and
Wetland Checklists

Sometimes we need to describe a variety of impacts on one particular eco-system or watershed. The following summaries and checklists can help you organize information to

- review basic wetland or stream conditions;
- describe wetland and stream features;
- provide a basis for asking questions and evaluating impacts; and
- identify a need for additional information (for example, advice from the appropriate expert).

The checklists can also guide natural resource protection efforts, restoration plans, or mitigation. Each checklist is followed by a summary of ecosystem benefits, checklist notes, and impact evaluation notes. This is a guide. You may want to add or delete features to customize your checklist.

Example 1: Assessing Stream Impacts

Stream Checklist

STREAM CHARACTERISTICS: SUMMARY FORM

Date:

Location:

Weather:

A. Watershed

Water source(s) (groundwater, precipitation, tributaries, lake, spring, or wetlands)	
Location of headwaters	
Connections (surface and underground waters)	
% impervious cover in subbasin	
% forested subbasin	
% wetlands in subbasin	
Presence of adjacent or floodplain wetlands	
Adjacent land use	
Floodplain (location, extent)	

B. Riparian buffer (for checklist purposes, consider a buffer of 300 feet from the water's edge)

Width of vegetated buffer	
Buffer vegetation type	
Buffer slope	
Percent of stream reach with vegetated buffer	

C. Bank and channel

Woody vegetation along bank	
Bank erosion	
Existing barriers between floodplain and channel	
Stream bottom (estimate percent rock, cobble, gravel, silt, mud, sand)	
Bottom structure (e.g., boulders, snags, debris)	

D. Water

Stream width and depth (with date)	
Stream flow in cfs (with date)	
Culvert discharges	
Temperature	
Turbidity	
Specific conductance	
Dissolved oxygen	
pH	
BOD	
Signs of contamination (odors, surface oils, off-color water)	
Runoff hot spots or point sources of pollution	
Additional data about specific contaminants	

E. Plants and animals (aquatic and riparian)

Group	Observed species	Possible species (based on region and habitat)
Plants		
Insects and other invertebrates		
Fish		
Amphibians		
Reptiles		
Birds		
Mammals		

F. Biodiversity

Nonnative invasive species	
Indicator species	
Species of conservation concern	
Development sensitive or tolerant	
Other notes (e.g., biomonitoring data, analysis, and level of stream impairment)*	

* Michael T. Barbour et al., *Rapid Bioassessment Protocols for Use in Streams and Wadeable Rivers: Periphyton, Benthic Macroinvertebrates and Fish*, 2nd ed., EPA 841-B-99-002 (Washington, DC: US Environmental Protection Agency, 1999).

Stream System Benefits

- Flood protection (e.g., slowing floodwaters, collecting runoff or floodwaters, and facilitating groundwater replenishment)
- Source of high-quality drinking water
- Nutrient cycling

- Downstream food-chain support
- Habitat (aquatic, riparian) and corridors
- Habitat for species of conservation concern or special interest
- Species for food, fiber, medicine, pharmaceuticals
- Water temperature moderation (e.g., plants provide shade that cools water)
- Source of water for agricultural use (e.g., irrigation, livestock watering)
- Tourism
- Recreation
- Quality of life and open space
- Education and research
- Economic value
- Natural control of stormwater runoff

Checklist Notes

List project activities and how they may affect stream characteristics and surrounding conditions. For example, land-use activities may have the following effects:

1. Reduce the size or extent of stream buffers, or their vegetation
2. Reduce forested cover in the subbasin
3. Increase impervious surfaces within the floodplain and subbasin
4. Reduce the extent of wetlands within the subbasin
5. Increase stormwater runoff and erosion, including new stormwater hot spots (e.g., agriculture, golf courses, feed lots, industrial use, mining, residential septic systems)
6. Increase sedimentation (e.g., deposit silt layer on stream bottom)
7. Erode or destabilize banks (e.g., grading, dredging, channelization, vegetation removal)
8. Change water flow, resulting in the following:
   - Too little water: loss of water source(s) or reduced groundwater supply, seasonal reduced flow, especially during drought; decreased normal in-stream flow
   - Too much water: increased water discharges and seasonal timing; increased normal in-stream flow
   - Altered timing of normal seasonal flows
9. Increase water consumption (water withdrawal, nearby groundwater withdrawal, irrigation and household water use), especially during drought
10. Change water chemistry by introducing high levels of road salt or other deicers applied on roads or accumulated in plowed snow piles within stream buffer

11. Introduce harmful chemicals in the form of herbicides, pesticides, and fungicides, as well as fertilizers within stream buffers (lawns, landscaping, agricultural fields, rights-of-way) or other areas that convey runoff into the stream
12. Change stormwater runoff (amount, location) and natural drainage patterns
13. Introduce point sources of pollution (e.g., culvert discharges, wastewater treatment, waste discharges)
14. Disturb stream bottom (e.g., deepening or dredging, or straightening stream channel)
15. Change the aquatic or riparian ecosystem:
    - Removing vegetation or changing vegetation type (e.g., replacing stream-bank shrubs with lawn)
    - Replacing native vegetation with nonnative species
    - Disrupting animal movement corridors along and within the stream
    - Disrupting or damaging habitat, food sources, breeding or nesting areas, and other features necessary to species survival
    - Altering water chemistry, which in turn affects aquatic habitat and organisms
    - Decreasing overall biodiversity of plants, invertebrates, fish, amphibians and reptiles, birds, and mammals
    - Introducing chemicals that accumulate in the food chain, harming fish-eating species (e.g., osprey or mink)
    - Changing predator-prey relationships
    - Changing the status of species that are indicators of specific aquatic conditions (e.g., trout that require high dissolved oxygen)
    - Creating conditions conducive to invasive nonnative plants and animals that may replace native species (e.g., knotweed along stream edges)
    - Decreasing overall biodiversity (changing the balance between habitat generalists and habitat specialists; increasing nonnative, invasive species; decreasing species of conservation concern as well as common native species)

Impact Evaluation Notes

1. Some species migrate along stream corridors. Species of conservation concern in downstream areas may move into upstream tributaries or vice versa when habitat conditions are favorable.

2. Waters within a subbasin are connected. Land-use impacts from one area of a watershed can affect downstream areas, groundwater, or other connected waters.
3. Consider the effects of all activities and sources of contamination, and describe cumulative impacts in terms of stream ecosystem health and watershed subbasin functions.
4. Use the information generated in this form to evaluate impacts on watershed and ecosystem services and benefits identified in the checklist.

Example 2: Assessing Wetland Impacts

Wetland Checklist

WETLAND CHARACTERISTICS: SUMMARY FORM

Date:

Location:

Weather:

A. Watershed

Watershed/subbasin	
Wetland boundary delineated	
Wetland size and type (e.g., marsh, hardwood swamp, meadow, vernal pool, fen, bog, prairie pothole)	
Size of contributing drainage area	
Connections (surface and underground waters)	
% impervious cover in contributing drainage area	
Adjacent land use	

B. Buffer (for checklist purposes, consider a buffer of 300 feet from the wetland's edge)

Width of vegetated buffer	
Buffer vegetation type	
Buffer slope	
% of wetland edge with vegetated buffer	
Source of erosion	

C. Water

Hydroperiod (based on wetland type)	
Water source(s) (stream or lake, spring, precipitation, groundwater)	
Water depth (multiple, seasonal measurements if possible, to document annual changes)	
Temperature	
Turbidity	
Specific conductance	
Dissolved oxygen	
pH	
BOD	
Signs of contamination (odors, surface oils, off-color water)	
Runoff hot spots or point sources of pollution	
Additional data about specific contaminants	

D. Plants and animals

Group	Observed species	Possible species (based on region and habitat)
Plants		
Insects and other invertebrates		
Fish		
Amphibians		

Reptiles		
Birds		
Mammals		

E. Biodiversity

Nonnative invasive species	
Indicator species	
Species of conservation concern	
Development sensitive or tolerant	
Other notes	

Wetland System Benefits

- Flood protection (storing floodwaters or runoff; slowing flood flows)
- Water-quality improvement
- Nutrient cycling
- Primary productivity and photosynthesis
- Habitat for plants and wildlife
- Habitat for species of conservation concern or special interest
- Species for food, fiber, medicine, pharmaceuticals
- Replenishment of groundwater
- Tourism
- Recreation
- Quality of life and open space
- Education and research
- Economic value
- Contribution to watershed health
- Natural control of stormwater runoff
- Erosion control
- Shoreline protection (lakes, rivers, streams)

- Protection and moderation of stream flows (contribute to more consistent stream flow, especially during drought)

Checklist Notes

List project activities and how they may affect wetland characteristics and surrounding conditions. For example, land-use activities may have the following effects:

1. Change wetland hydroperiod, resulting in
   - too little water; decreased water depth especially during drought; less water available for groundwater or stream replenishment;
   - too much water; increased water depth and its seasonal timing; reduced wetland storage capacity;
   - altered seasonal water depth (even small changes influence which wetland plants grow and thrive in a wetland and its buffer).
2. Deplete wetland water via increased consumption (water withdrawal, groundwater withdrawal, and household water use)
3. Interfere with groundwater-wetland connection and water supply
4. Change the flow of water into, through, and from the wetland
5. Alter land use within 300 feet of wetland, or within its contributing drainage area, leading to increased contamination from stormwater runoff
6. Introduce point sources of pollution, including culvert discharges
7. Increase the percent of impervious surface within the contributing drainage area
8. Include application of herbicides, pesticides, and fungicides, or fertilizers within wetland or buffer (lawns, landscaping, agricultural fields, rights-of-way)
9. Reduce the size or extent of the wetland buffer
10. Degrade buffer vegetation (introduction of invasive species, natural vegetation in poor condition)
11. Erode wetland edges (e.g., grading, vegetation removal)
12. Dredge or fill a wetland
13. Dump wastes within wetland (yard waste, trash, debris)
14. Include stormwater management-induced changes in wetland water supply and hydroperiod that divert water away from wetland or discharge stormwater runoff into wetland

15. Create stormwater management ponds and other facilities that store or discharge water within 100 feet of wetland
16. Create water-flow constrictions (road crossings, culverts) within the wetland and between wetlands that are hydrologically connected
17. Alter water chemistry
18. In cold climates, use road salt within 300 feet of water (roads and other paved surfaces) or deposit plowed snow into wetland
19. Change the wetland ecosystem:
    - Removing vegetation, or changing vegetation type (e.g., replacing wetland edge bushes or sedges with lawn)
    - Replacing native vegetation with nonnative species
    - Disrupting animal movement corridors
    - Disrupting or damaging habitat, food sources, breeding or nesting areas, and other features necessary to species survival (e.g., turtle nesting areas, aquatic edge vegetation for sheltering larval stages)
    - Reducing or degrading critical terrestrial habitat for vernal-pool-breeding amphibians (usually a wooded area that extends approximately 700 feet from the edge of a vernal pool)[1]
    - Altering water chemistry, which in turn affects aquatic habitat and organisms
    - Decreasing overall biodiversity of plants, invertebrates, fish, amphibians and reptiles, birds, and mammals
    - Introducing chemicals that accumulate in the food chain, harming fish-eating species (e.g., osprey or mink)
    - Changing predator-prey relationships
    - Changing the status of species that are indicators of specific habitat conditions
    - Creating conditions conducive to invasive nonnative plants and animals that may replace native species (e.g., knotweed along stream edges)
    - Decreasing overall biodiversity; changing the balance between habitat generalists and habitat specialists; increasing nonnative, invasive plant or animal species; decreasing species of conservation concern and common native species

Impact Evaluation

1. Identify the ecosystem benefits provided by this wetland and their value to your community.
2. Identify the role of the wetland within the watershed or subbasin (e.g., water storage).

3. Add together all impacts to this wetland, its buffer, and its contributing drainage area from all sources and describe cumulative impacts in terms of wetland health.
4. Identify interconnections with other wetlands, streams, ponds and lakes, and groundwater. Most wetlands don't have "downstream areas" unless they are directly connected to a stream system, but many of them intersect the water table and have direct contact with groundwater.
5. Use the information generated in this form to evaluate impacts on ecosystem services and benefits.

notes

Introduction

1. Richard Harwood, "Earth Day Stirs the Nation," *Washington Post*, April 23, 1970.
2. Maine Department of Inland Fisheries and Game, *Fish Management in the Kennebec River*, 1969, 40.
3. Daniel Michor, "People in Nature: Environmental History of the Kennebec River, Maine" (master's thesis, University of Maine, 2003).
4. "Current High Volume Horizontal Hydraulic Fracturing Drilling Bans and Moratoria in NY State," *Fractracker Alliance*, November 22, 2013, http://www.fractracker.org/map/ny-moratoria/.
5. *New York Times*, December 17, 2014, http://www.nytimes.com/2014/12/18/nyregion/cuomo-to-ban-fracking-in-new-york-state-citing-health-risks.
6. *USA Today*, October 13, 2013, http://www.usatoday.com/story/news/nation/2013/10/13/lake-erie-algae-drinking-water/2976273/.
7. *New York Times*, July 24, 2013, http://www.nytimes.com/2013/07/25/us/louisiana-agency-to-sue-energy-companies-for-wetland-damage.
8. *New York Times*, May 19, 2013, http://www.nytimes.com/2013/05/20/us/high-plains-aquifer-dwindles-hurting-farmers.html.
9. Andrew Blaustein, John M. Romansic, Joseph M. Kiesecker, and Audrey C. Hatch, "Ultraviolet Radiation, Toxic Chemicals, and Amphibian Population Declines," *Diversity and Distributions* 9 (2003): 123–40.
10. Fred Pearce, *Downstream Voices: Wetland Solutions to Reducing Disaster Risk* (The Netherlands: Wetlands International, 2014).
11. "Wetlands: Status and Trends," US Environmental Protection Agency, "http://water.epa.gov/type/wetlands/vital_status.cfm.
12. K. Averyt et al., "Sectoral Contributions to Surface Water Stress in the Coterminous United States," *Environmental Research Letters* 8, no. 3 (2013).
13. Zoological Society of London and Global Footprint Network, *2010 and Beyond: Rising to the Biodiversity Challenge*, World Wildlife Fund, 2004, http://www.wwf.org.uk/filelibrary/pdf/2010_and_beyond.pdf.
14. Reed F. Noss, Edward LaRoe, and J. Michael Scott, *Endangered Ecosystems of the United States: A Preliminary Assessment of Loss and Degradation*, 1995, http://biology.usgs.gov/pubs/ecosys.htm.
15. Carlos Corvalan et al., *Millennium Ecosystem Assessment: Ecosystems and Human Well-Being; Synthesis* (Geneva, Switzerland: World Health Organization, 2005).
16. Joseph Guth, "Cumulative Impacts: Death-Knell for Cost-Benefit Analysis in Environmental Decisions, *Barry Law Review* 11 (Fall 2008).
17. Ibid., 23.

18. Bob Deans, *Reckless: The Political Assault on the American Environment* (Lanham, MD: Rowman & Littlefield, 2012), 1.

19. The Clean Water Act (CWA) establishes the basic structure for regulating discharges of pollutants into the waters of the United States and regulating quality standards for surface waters. The basis of the CWA was enacted in 1948 and was called the Federal Water Pollution Control Act, but the act was significantly reorganized and expanded in 1972 and amended in 1977 and 1987. The "Clean Water Act" became the act's common name. Under the CWA, EPA has implemented pollution control programs such as setting wastewater standards for industry and setting water-quality standards for all contaminants in surface waters. The Clean Water Act does not directly address groundwater contamination, which is included in the Safe Drinking Water Act, the Resource Conservation and Recovery Act, and the Superfund Act.

20. Supreme Court of the United States, *Ohio Valley Environmental Coalition, et al., v. United States Army Corps of Engineers, et al.*, August 2009, http://earthjustice.org/sites/default/files/library/legal_docs/final_mtr_cert_petition_08-2009.pdf.

21. John McQuaid, "Mining the Mountain," *Smithsonian*, January 2009.

22. Adam Isen, Maya Rossin-Slater, and W. Reed Walker, "Every Breath You Take—Every Dollar You'll Make: The Long-Term Consequences of the Clean Air Act of 1970," National Bureau of Economic Research Working Paper No. 19858, January 2014, http://www.nber.org/papers/w19858.pdf?new_window=1.

23. Martin Spray, "Entire Economy Depends on Healthy Environment, Report Shows," *Guardian*, November 19, 2013.

24. Amy Westervelt, "Better Profits through Green Chemistry," *Forbes*, December 28, 2011, http://www.forbes.com/sites/amywestervelt/2011/12/28/better-profits-through-green-chemistry/.

25. "Creating Equitable, Healthy, and Sustainable Communities: Strategies for Advancing Smart Growth, Environmental Justice and Equitable Development," EPA 231-K-10-005, February 2013, http://www.epa.gov/smartgrowth/pdf/equitable-dev/equitable-development-report-508-011713b.pdf.

26. Kyriaki Remoundou and Phoebe Koundouri, "Environmental Effects on Public Health: An Economic Perspective," *International Journal of Environmental Research and Public Health* 6 (2009): 2160–78.

27. Donald F. Harker and Elizabeth U. Natter, *Where We Live: A Citizen's Guide to Conducting a Community Environmental Inventory* (Washington, DC: Island Press, 1995), 2.

28. Millennium Ecosystem Assessment Board, *Living beyond Our Means: Natural Assets and Human Well-Being*, United Nations Millennium Ecosystem Assessment, 2005, p. 5, http://www.maweb.org/en/BoardStatement.aspx.

29. The national survey of voters was conducted by the bipartisan research team of Fairbank, Maslin, Maullin, Metz & Associates (D) and Public Opinion Strategies (R), June 16–19, at the request of the Nature Conservancy. See more at http://www.nature.org/newsfeatures/pressreleases/poll-conservation-is-patriotic-and-has-bipartisan-support.xml#sthash.VoTRFUoi.dpuf.

30. "A Look into the Ocean's Future," editorial, *New York Times*, July 15, 2011.

31. Amitai Etzioni, *The Spirit of Community: The Reinvention of American Society* (New York: Simon & Schuster, 1993), 11.

1. Natural Water Systems and Their Benefits

Epigraph quote from "Ecological Stewardship and Economic Development: Do We Have to Choose?," presentation at the annual meeting for the Connecticut Association of Wetland Scientists, New Haven, CT, February 23, 2011.

1. "What on Earth Do You Know about Water?" US Environmental Protection Agency, updated February 24, 2011, http://www.epa.gov/gmpo/edresources/water_5.html.

2. "How Many Baths Could I Get from a Rainstorm?," US Geological Survey, http://ga.water.usgs.gov/edu/qa-home-baths.html.

3. Mark Hebblewhite et al., "Human Activity Mediates a Trophic Cascade Caused by Wolves," *Ecology* 86 (2005): 2135–44.

4. *White-Nose Syndrome*, National Wildlife Health Center, USGS, http://www.nwhc.usgs.gov/disease_information/white-nose_syndrome/.

5. David Pimentel, Rudolfo Zuniga, and Doug Morrison, "Update on the Environmental and Economic Costs Associated with Alien-Invasive Species in the United States," *Ecological Economics* 52 (2005): 273–88.

6. John Rothlisberger et al., "Ship-Borne Nonindigenous Species Diminish Great Lakes Ecosystem Services," *Ecosystems* 15 (2012): 1–15.

7. "The Economic Cost of Large Constrictor Snakes," US Fish and Wildlife Service, January 2012, http://www.fws.gov/home/feature/2012/pdfs/EconImpact.pdf.

8. "The Cost of Invasive Species," US Fish and Wildlife Service, January 2012, http://www.fws.gov/home/feature/2012/pdfs/costofinvasivesfactsheet.pdf.

9. Walter Reid et al., *Millennium Ecosystem Assessment: Ecosystems and Human Well-Being Synthesis* (Washington, DC: Island Press, 2005).

10. T. Williams, "What Good Is a Wetland?," *Audubon* 98 (1996).

11. "Did You Know . . . Healthy Wetlands Devour Mosquitoes," Indiana Department of Natural Resources: Indiana Wetlands Conservation Plan Fact Sheet, http://www.in.gov/dnr/fishwild/files/hlywet.pdf.

12. Richard S. Ostfeld and Felicia Keesing, "Biodiversity and Disease Risk: The Case of Lyme Disease," *Conservation Biology* 14 (June 2000).

13. USDA Natural Resources Conservation Service, "Native Pollinators," Fish and Wildlife Habitat Management Leaflet no. 34.2005, http://plants.usda.gov/pollinators/Native_Pollinators.pdf.

14. Janine Benyus, *Biomimicry: Innovation Inspired by Nature* (New York: William Morrow and Co., 1997).

15. "Biomimicry," NYSERDA, www.nyserda.ny.gov/Energy-Innovation-and-Business-Development/Research-and-Development/Biomimicry.aspx.

16. Biomimicry Institute, http://www.biomimicryinstitute.org.

17. US Environmental Protection Agency, "Wetland Functions and Values," Watershed Academy Web, http://cfpub.epa.gov/watertrain/pdf/modules/WetlandsFunctions.pdf.

18. US Environmental Protection Agency, "Wetlands: Protecting Life and Property from Flooding," EPA843-F-06-001, 2006, http://water.epa.gov/type/wetlands/outreach/upload/Flooding.pdf.

19. David Batker et al., *Gaining Ground: Wetlands, Hurricanes, and the Economy: The Value of Restoring the Mississippi River Delta* (Tacoma, WA: Earth Economics, 2010).

20. "Water: Rivers, and Streams," US Environmental Protection Agency, 2014, http://water.epa.gov/type/rsl/streams.cfm#types.

21. Lewis M. Cowardin et al., *Classification of Wetlands and Deepwater Habitats of the United States*, US Fish and Wildlife Service, FWS/OBS-79/31 (Washington, DC: Government Printing Office,1979), 3.

22. Ibid.

23. National Wetland Inventory, US Fish and Wildlife Service, http://www.fws.gov/wetlands/Data/Wetland-Codes.html.

24. R. A. Gleason et al., *Estimating Water Storage Capacity of Existing and Potentially Restorable Wetland Depressions in a Subbasin of the Red River of the North*, US Geological Survey Open-File Report 2007–1159, 2007.

25. South Carolina Department of Health and Environmental Control, comment letter on the Advanced Notice of Proposed Rulemaking on the Clean Water Act Regulatory Definition of "Waters of the United States," docket ID: EPA-HQ-OW-2002–0050–1337, 2003.

26. "Wetlands and Water Quality," Purdue University Cooperative Extension, www.extension. purdue.edu/extmedia/WQ/WQ-10.html.

27. Paul Barlow and Stanley Leake, *Streamflow Depletion by Wells—Understanding and Managing the Effects of Groundwater Pumping on Streamflow*, US Geological Survey Circular 1376 (Reston, VA: Government Printing Office, 2012), http://pubs.usgs.gov/circ/1376/.

28. William Alley, T. Reilly, and O. Franke, *Sustainability of Ground-Water Resources*, US Geological Survey Circular 1186 (Reston, VA: Government Printing Office, 1999), http://pubs.usgs. gov/circ/circ1186/html/gen_facts.html.

29. C. Ernst, R. Gullick, and K. Nixon, "Protecting the Source: Conserving Forests to Protect Water," *Opflow* 30 (2004).

30. David Batker et al., *The Puyallup River Watershed: An Ecological Economic Characterization* (Tacoma, WA: Earth Economics, 2011), www.eartheconomics.org.

31. Ibid., 2.

2. Picturing Environmental Features and Systems

1. Karen Cappiella et al., *Adapting Watershed Tools to Protect Wetlands* (Ellicott City, MD: Center for Watershed Protection, 2005), 6–9.

2. US Army Corps of Engineers, National Wetland Plant List home site, http://rsgisias.crrel. usace.army.mil/NWPL/.

3. Elizabeth Johnson and Michael Klemens, "The Impacts of Sprawl on Biodiversity," in *Nature in Fragments: The Legacy of Sprawl*, ed. Elizabeth Johnson and Michael Klemens (New York: Columbia University Press, 2005), 22–30.

4. "2006 Land Cover Data," US Environmental Protection Agency, http://www.epa.gov/mrlc/ nlcd-2006.html.

5. Gregory Edinger et al., *Ecological Communities of New York State* (Albany: New York State Department of Environmental Conservation, 2002).

6. Erik Kiviat and Gretchen Stevens, *Biodiversity Assessment Manual for the Hudson River Estuary Corridor* (Annandale, NY: Hudsonia Ltd., 2001), 200.

7. Scott J. Goetz et al., "IKONOS Imagery for Resources Management: Tree Cover, Impervious Surfaces, and Riparian Buffer Analysis in the Mid-Atlantic Region," *Remote Sensing of Environment* 88 (2003): 195–208.

8. Linda Exum et al., *Estimating and Projecting Impervious Cover in the Southeastern United States*, EPA/600/R-05/061 (Athens, GA: National Exposure Research Laboratory, US Environmental Protection Agency, 2005).

9. Thomas Schueler, *Controlling Urban Runoff: A Practical Manual for Planning and Designing Urban BMP's* (Washington DC: Metropolitan Information Center, 1987), http://www.dec. ny.gov/docs/water_pdf/simple.pdf.

10. Edward O. Wilson, "Within One Cubic Foot," *National Geographic*, February 2010.

11. Russell Urban-Mead, *Dutchess County Aquifer Recharge Rates and Sustainable Septic System Density Recommendations* (Poughkeepsie, NY: Chazen Cos., 2006), 34.

12. Christina Kennedy, Jessica Wilkinson, and Jennifer Balch, *Conservation Thresholds for Land Use Planners* (Washington, DC: Environmental Law Institute, 2003), 21–23.

13. Ibid., 20.

14. Lynn Boyd, *Wildlife Use of Wetland Buffer Zones and Their Protection under the Massachusetts Wetland Protection Act* (Boston: University of Massachusetts, Department of Natural Resources Conservation, 2001).

15. Derek Booth, "Forest Cover, Impervious-Surface Area, and the Mitigation of Urbanization Impacts in King County, Washington," University of Washington Department of Civil and Environmental Engineering, 2000, http://hdl.handle.net/1773/19552.

16. Goetz et al., "IKONOS Imagery," 195–208.

17. Caryn Ernst, *Protecting the Source* (San Francisco: Trust for Public Land, 2004), 21.

18. Gary Bentrup, *Conservation Buffers: Design Guidelines for Buffers, Corridors, and Greenways*, Gen. Tech. Rep.SRS-109 (Asheville, NC: Department of Agriculture, Forest Service, Southern Research Station, 2008), 110.

19. James McElfish, Rebecca Kihslinger, and Sandra Nichols, *Planner's Guide to Wetland Buffers for Local Governments* (Washington, DC: Environmental Law Institute, 2008), 25.

20. Paul Mayer, Steven Reynolds, and Timothy Canfield, *Riparian Buffer Width, Vegetative Cover, and Nitrogen Removal Effectiveness: A Review of Current Science and Regulations*, EPA/600/R-05/118 (Ada, OK: US Environmental Protection Agency, 2005).

21. Ralph Tiner, *Wetland Indicators: A Guide to Wetland Identification, Delineation, Classification, and Mapping* (Boca Raton, FL: CRC Press, 1999), 33.

22. US Army Corps of Engineers, National Wetland Plant List home site, http://rsgisias.crrel.usace.army.mil/NWPL/.

23. Tiner, *Wetland Indicators*, 103.

24. Ibid., 71–87.

25. Michael T. Barbour et al., *Rapid Bioassessment Protocols for Use in Streams and Wadeable Rivers: Periphyton, Benthic Macroinvertebrates, and Fish*, 2nd ed., EPA 841-B-99–002 (Washington, DC: US Environmental Protection Agency, 1999).

26. Roxanne Thomas et al., *State Wetland Protection: Status, Trends, and Model Approaches* (Washington, DC: Environmental Law Institute, 2008), 67.

27. All fifty states have adopted a state wildlife action plan, as authorized and funded by the US Congress. For details see http://wildlifeactionplans.org/.

28. Aram Calhoun and Michael W. Klemens, *Best Development Practices: Conserving Pool-Breeding Amphibians in Residential and Commercial Developments in the Northeastern United States*, MCA Technical Paper no. 5 (New York: Wildlife Conservation Society, 2002).

3. How Land-Use Activities Affect Water

1. Doug Pflugh, "Our Nation's Most Endangered River—the Colorado," Earthjustice blog, April 17, 2013, http://earthjustice.org/blog/2013-april/our-nation-s-most-endangered-river-the-colorado.

2. "Chesapeake Bay Watershed," Chesapeake Bay Program, http://www.chesapeakebay.net/discover/baywatershed.

3. William Baker, "President's Message," in *2012 State of the Bay Report* (Annapolis, MD: Chesapeake Bay Foundation, 2012), 2, http://www.cbf.org/about-the-bay/state-of-the-bay/2012-report.

4. Laura Gottesdiener, "Mississippi River Flooding Reaching Historic Levels," *Huffington Post*, updated July 11, 2011, http://www.huffingtonpost.com/2011/05/11/mississippi-flooding-river-historic_n_860835.html.

5. Russell Urban-Mead, *Dutchess County Aquifer Recharge Rates and Sustainable Septic System Density Recommendations* (Poughkeepsie, NY: Chazen Cos., 2006), 34.

6. Michelle Hladik, Dana Kolpin, and Kathryn Kuivila, "Widespread Occurrence of Neonicotinoid Insecticides in Streams in a High Corn and Soybean Producing Region, USA," *Environmental Pollution* 193 (October 2014): 189–96, http://www.sciencedirect.com/science/article/pii/S0269749114002802.

7. Paul Thacker, "American Lawns Impact Nutrient Cycles," *Environmental Science and Technology* (2005): 98, http://pubs.acs.org/doi/pdf/10.1021/es053200p.

8. Thomas Schueler, "Urban Pesticides: From the Lawn to the Stream," *Watershed Protection Techniques* (1995): 247–53.

9. Bill Chameides, "Statistically Speaking: Lawns by the Numbers," Nicholas School of the Environment at Duke University, Durham, NC, July 25, 2008, https://blogs.nicholas.duke.edu/thegreengrok/lawns_stats/.

10. "Pets by the Numbers," Humane Society, January 20, 2014, http://www.humanesociety.org/issues/pet_overpopulation/facts/pet_ownership_statistics.html#.UzBQrVyBX1o.

11. Diana M. Foster, "Who Knew? Upcycling the Dog Poo," *New York Times Green Blog*, April 4, 2012, http://green.blogs.nytimes.com/2012/04/04/who-knew-upcycling-the-dog-poo/?_php=true&_type=blogs&_r=0.

12. "Source Water Protection Practices Bulletin: Managing Pet and Wildlife Waste to Prevent Contamination of Drinking Water," US Environmental Protection Agency, EPA 916-F-01–027, July 2001, http://www.epa.gov/safewater/sourcewater/pubs/fs_swpp_petwaste.pdf.http://cfpub.epa.gov/npdes/stormwater/menuofbmps/index.cfm?action=browse&Rbutton=detail&bmp=4.

13. Ibid.

14. "Pesticide News Story: EPA Releases Report Containing Latest Estimates of Pesticide Use in the United States," US Environmental Protection Agency, February 17, 2011, http://epa.gov/oppfead1/cb/csb_page/updates/2011/sales-usage06–07.html.

15. Maya Van Rossum, "Roundup Chemical Found in Breast Milk," Safer Chemicals, Healthy Families, May 21, 2014, http://saferchemicals.org/2014/05/21/roundup-chemical-found-in-breast-milk/.

16. Thomas Dahl, Craig Johnson, and W. E. Frayer, *Wetlands: Status and Trends in the Conterminous United States, Mid-1970's to Mid 1980's* (Washington, DC: US Department of the Interior, Fish and Wildlife Service, 1991), 2.

4. Measuring the Impacts

1. John Flesher, "Erie Algae: Report Says Toxic Mega-Blooms Could Become the New Norm," *Pittsburgh Post-Gazette*, October 9, 2013.

2. Sharon Behar, *Testing the Waters: Chemical and Physical Vital Signs of a River* (Montpelier, VT: River Watch Network, 1997).

3. "Introduction to Sediment and River Stability," US Environmental Protection Agency, http://water.epa.gov/scitech/datait/tools/warsss/sedsource_index.cfm.

4. "Turbidity: Why Is It Important?," Water on the Web, University of Minnesota, January 17, 2008, www.waterontheweb.org/under/waterquality/turbidity.html.

5 Charles P. Newcombe and Jorgen O. T. Jensen, "Channel Suspended Sediment and Fisheries: A Synthesis for Quantitative Assessment of Risk and Impact," *North American Journal of Fisheries Management* 16 (1996): 693–727.

6. Anne Bernhard, "The Nitrogen Cycle: Processes, Players, and Human Impact," *Nature Education Knowledge* 3 (2012): 25; Nature Education Knowledge Project, "The Nitrogen Cycle: Processes, Players, and Human Impact," http://www.nature.com/scitable/knowledge/library/the-nitrogen-cycle-processes-players-and-human-15644632.

7. Thomas Schueler, Lisa Fraley-McNeal, and Karen Cappiella, "Is Impervious Cover Still Important? Review of Recent Research," *Journal of Hydrologic Engineering* 14 (2009): 309.

8. "Implications of the Impervious Cover Model: Stream Classification, Urban Subwatershed Management and Permitting," Chesapeake Stormwater Network, CSN Technical Bulletin no. 3 (2008), http://chesapeakestormwater.net/wp-content/uploads/downloads/2012/01/CSN20TB20No20320The20ICM1.pdf.

9. Tiffany Wright et al., *Direct and Indirect Effects of Urbanization on Wetland Quality* (Ellicott City, MD: Center for Watershed Protection, 2006).

10. "Toxic Flood: Why We Need Stronger Regulations to Protect Public Health from Industrial Water Pollution," Food and Water Watch and the Political Economy Research Institute, May 16, 2013, http://www.foodandwaterwatch.org/reports/a-toxic-flood/.

11. Mark Fischetti, "The Great Chemical Unknown: A Graphical View of Limited Lab Testing," *Scientific American*, November 2010, http://www.scientificamerican.com/article.cfm?id=the-great-chemical-unknown.

12. "Emerging Chemicals of Concern," California Department of Toxic Substances Control, http://www.dtsc.ca.gov/assessingrisk/emergingcontaminants.cfm.

13. "Chemical Testing and Data Collection," US Environmental Protection Agency, http://www.epa.gov/opptintr/chemtest/pubs/mtlintro.html.

14. "Impacts of Pharmaceuticals and Personal Care Products," American Rivers, http://www.americanrivers.org/initiatives/pollution/pharmaceuticals/#.

15. Ibid.

16. "Drinking Water Contaminants: List of Contaminants and Their MCLs," US Environmental Protection Agency, June 3, 2013, http://water.epa.gov/drink/contaminants/#List.

17. Ibid.

18. Sandra Steingraber, *Living Downstream: A Scientist's Personal Investigation of Cancer and the Environment* (New York: Vintage Books, 1998), 194.

19. "Endocrine Disruption," Endocrine Disruption Exchange (TEDX), 2014, http://endocrinedisruption.org/endocrine-disruption/introduction/overview.

20. Renee Sharp and J. Paul Pestano, *Water Treatment Contaminants: Forgotten Toxics in American Water* (Washington, DC: Environmental Working Group, February 2013), http://static.ewg.org/reports/2013/water_filters/2013_tap_water_report_final.pdf.

21. Douglas McIntyre, "10 US Cities with the Worst Drinking Water," Going Green, NBC News, February 3, 2011, http://www.nbcnews.com/id/41354370/ns/business-going_green/t/us-cities-worst-drinking-water/#.

22. "Residential Water Testing," National Testing Laboratories, http://www.ntllabs.com/residential.html.

23. "NFIP Statistics," National Flood Insurance Program, updated March 28, 2014, https://www.floodsmart.gov/floodsmart/pages/media_resources/stats.jsp.

24. Ibid.

25. US General Accountability Office, *Water Quality: Better Data and Evaluation of Urban Runoff Programs Needed to Address Effectiveness*, GAO-01-679, Washington, DC, 2001.

26. Ibid., 16.

27. Michael J. Paul and Judy L. Meyer, "The Ecology of Urban Streams," *Annual Review of Ecology and Systematics* 32 (2001): 333–65.

28. "CADDIS Volume 2: Sources, Stressors and Responses," US Environmental Protection Agency, July 31, 2012, http://www.epa.gov/caddis/ssr_urb_is4.html.

29. Thomas Schueler, "The Importance of Imperviousness," *Watershed Protection Techniques* 1 (1994): 100–111.

30. Thomas E. Dahl, *Wetlands Losses in the United States, 1780's to 1980's* (Washington, DC: US Department of the Interior, Fish and Wildlife Service, 1990), http://www.npwrc.usgs.gov/resource/wetlands/wetloss/.

31. Wright et al., *Direct and Indirect Effects of Urbanization on Wetland Quality*.

32. "Blood Vessels and Circulation," University of Miami, FL, http://www.as.miami.edu/chemistry/2086/Chapter_21/NEW-Chap21_class_part1.htm.

33. Christina Kennedy, Jessica Wilkinson, and Jennifer Balch, *Conservation Thresholds for Land Use Planners* (Washington, DC: Environmental Law Institute, 2003), 19.

34. "Watershed Forestry Resource Guide," Center for Watershed Protection and US Forest Service–Northeastern Area State & Private Forestry, 2008, http://www.forestsforwatersheds.org/urban-watershed-forestry/.

35. Ibid.

36. Gary Bentrup, *Conservation Buffers: Design Guidelines for Buffers, Corridors, and Greenways*, Gen. Tech. Rep. SRS-109 (Asheville, NC: US Department of Agriculture, Forest Service, 2008).

37. William M. Ally, Thomas Reilly, and O. Lehn Franke, *Sustainability of Ground-Water Resources*, US Geological Survey Circular 1186 (Denver: US Geological Survey, 1999), 15.

38. Ibid., 19.

39. Ibid., 79.

40. Andrew Steer, "Water Risk on the Rise," World Resources Institute, September 20, 2013, http://www.wri.org/blog/water-risk-rise.

41. Francis Gassert et al., "Aqueduct Global Maps 2.0," working paper, World Resources Institute, 2013, http://wri.org/publication/aqueduct-global-maps-20.

42. Bernd Blossey et al., "Impact and Management of Purple Loosestrife (*Lythrum salicaria*) in North America," *Biodiversity and Conservation* 10 (2001): 1787–1807.

43. Douglas Tallamy, *Bringing Nature Home* (Portland, OR: Timber Press, 2007), 358.

44. Ibid., 329.

45. Ibid., 147.

46. Danielle LaBruna and Michael W. Klemens, *Northern Wallkill Biodiversity Plan: Balancing Development and Environmental Stewardship in the Hudson River Valley Watershed*, MCA Technical Paper 13 (New York: Wildlife Conservation Society, 2007).

47. Ibid.

5. Energy Development and Water

1. Kevin Liptak, "Keystone Delay a Political Move, Republicans Say," CNN, December 17, 2011, http://www.kpax.com/news/keystone-delay-a-political-move-republicans-say/.

2. "Say No to the Keystone XL," editorial, *New York Times*, October 2, 2011, http://www.nytimes.com/2011/10/03/opinion/say-no-to-the-keystone-xl.html?_r=0.

3. Emily Weidenhof, Vivian Ngo, and Richard Gonzalez, *Hancock and the Marcellus Shale: Visioning the Impacts of Natural Gas Extraction along the Upper Delaware* (New York: Open Space Institute, 2009).

4. Nels Johnson et al., *Pennsylvania Energy Impacts Assessment, Report 1: Marcellus Shale Natural Gas and Wind* (Harrisburg, PA: Nature Conservancy, 2010).

5. Kim Martineau, "Ohio Quakes Probably Triggered by Disposal Well, Say Seismologists," Earth Institute, Columbia University, January 6, 2012, http://blogs.ei.columbia.edu/2012/01/06/seismologists-link-ohio-earthquakes-to-waste-disposal-wells/.

6. Ted Fink, *Middlefield Land Use Analysis: Heavy Industry and Oil, Gas or Solution Mining and Drilling* (Rhinebeck, NY: Greenplan Inc., 2011), 31.

7. Monika Freyman and Ryan Salmon, *Hydraulic Fracturing and Water Stress: Growing Competitive Pressures for Water* (Boston: Ceres, 2013), 5.

8. Weidenhof, *Hancock and the Marcellus Shale.*

9. Mary B. Adams et al., *Effects of Development of a Natural Gas Well and Associated Pipeline on the Natural and Scientific Resources of the Fernow Experimental Forest*, US Department of Agriculture, Forest Service: Gen. Tech. Rep. NRS-76, 2011.

10. Craig Michaels, James Simpson, and William Wegner, *Fractured Communities: Case Studies of the Environmental Impacts of Industrial Gas Drilling* (New York: Riverkeeper, 2010).

11. Ibid.

12. Fink, *Middlefield Land Use Analysis.*

13. Committee on Energy and Commerce, *Chemicals Used in Hydraulic Fracturing* (Washington DC: US House of Representatives, April 2011), http://democrats.energycommerce.house.gov/sites/default/files/documents/Hydraulic-Fracturing-Chemicals-2011-4-18.pdf.

14. Krishna Ramanujan, "Study Suggests Hydrofracking Is Killing Farm Animals, Pets," *Cornell Chronicle*, March 7, 2012, http://www.news.cornell.edu/stories/2012/03/reproductive-problems-death-animals-exposed-fracking.

15. Stephen Osborn et al., "Methane Contamination of Drinking Water Accompanying Gas-Well Drilling and Hydraulic Fracturing," *Proceedings of the National Academy of Sciences* 108 (2011): 8172–76, www.pnas.org/cgi/doi/10.1073/pnas.1100682108.

16. "The TEDX List of Potential Endocrine Disruptors," Endocrine Disruption Exchange, 2014, http://endocrinedisruption.org/endocrine-disruption/tedx-list-of-potential-endocrine-disruptors/overview.

17. E. L. Rowan, *Radium Content of Oil and Gas Field Produced Waters of Northern Appalachian Basin (USA)—Summary and Discussion of Data*, US Geological Survey Scientific Investigations Report 2011–5135 (Reston, VA: USGS, 2011), http://water.epa.gov/lawsregs/rulesregs/sdwa/radionuclides/basicinformation.cfm.

18. Sheila Bushkin-Bedient, Geoffrey Moore, and the Preventive Medicine and Family Health Committee of the Medical Society of the State of New York, "Update on Hydrofracking," NYS American Academy of Pediatrics, http://nysaap.org/update-on-hydrofracking/.

19. "TEDX List."

20. "Water Quality and Gas Well Drilling," WaterCheck, http://www.watercheck.com/gaswelldrilling.html.

21. Bruce Finley, "Benzene Levels in Parachute Creek near Gas Plant Spill Double Again," *Denver Post*, July 18, 2013, http://www.denverpost.com/ci_23689092/parachute-creek-benzene-levels-double-again.

22. John Upton, "Fracking Accident Leaks Benzene into Colorado Stream," *Terra News*, May 16, 2013, http://www.theterranews.com/content/?p=59452.

23. Monique Beaudin, "Lac-Mégantic Disaster: Where Things Stand Today," *Montreal Gazette*, January 23, 2014, http://www.montrealgazette.com/news/Mégantic+disaster+Where+things+stand+today/9418300/story.html.

24. US Department of State, *Final Supplemental Environmental Impact Statement for the Keystone XL Project Executive Summary* (Washington, DC: US Department of State, January 2014), http://keystonepipeline-xl.state.gov/documents/organization/221135.pdf.

25. Tara Thean, "Why the Yellowstone Spill Is So Tough to Clean Up," *Time Science and Space*, July 11, 2011, http://science.time.com/2011/07/11/why-the-yellowstone-oil-spill-is-so-tough-to-clean-up/#ixzz2Ws0fclMo.

26. "Tox Faqs," Agency for Toxic Substances and Disease Registry, March 3, 2011, http://www.atsdr.cdc.gov/toxfaqs/index.asp.

27. US Environmental Protection Agency, letter to Enbridge Energy, March 14, 2013, http://www.documentcloud.org/documents/619981-epa-kalamazoo-order-letter-mar-14-2013.html#document/p5/a9579.

28. Lisa Song, "Cleanup of 2010 Michigan Dilbit Spill Aims to Stop Spread of Submerged Oil," *Inside Climate News*, March 27, 2013, http://insideclimatenews.org/news/20130327/cleanup-2010-mich-dilbit-spill-aims-stop-spread-submerged-oil.

29. David Biello, "Does Tar Sand Oil Increase the Risk of Pipeline Spills?," *Scientific American*, April 4, 2013, http://www.scientificamerican.com/article.cfm?id=tar-sand-oil-and-pipeline-spill-risk.

30. Thean, "Why the Yellowstone Spill Is So Tough to Clean Up."

31. John P. Incardona et al., "Deepwater Horizon Crude Oil Impacts the Developing Hearts of Large Predatory Pelagic Fish," *Proceedings of the National Academy of Sciences*, February 24, 2014, http://www.pnas.org/content/early/2014/03/19/1320950111.

32. B. E. Ballachey et al., *2013 Update on Sea Otter Studies to Assess Recovery from the 1989 Exxon Valdez Oil Spill, Prince William Sound, Alaska*, US Geological Survey Open-File Report 2014–1030, February 2014, http://pubs.usgs.gov/of/2014/1030/.

33. "Coal Production in the United States—an Historical Overview," US Energy Information Administration, 2006, http://www.eia.gov/coal/.

34. "Extent of Mountaintop Mining in Appalachia," ILoveMountains, http://ilovemountains.org/reclamation-fail/details.php.

35. US Environmental Protection Agency, *Mountaintop Mining / Valley Fills in Appalachia: Final Programmatic EIS*, EPA 9–03-R-05002 (Philadelphia: US Environmental Protection Agency,

2005), http://nepis.epa.gov/Exe/ZyNET.exe/20005XA6.TXT?ZyActionD=ZyDocument&Client=EPA&Index=2000+Thru+2005&Docs=&Quer.

36. "Mountaintop Mining, West Virginia," NASA Earth Observatory, http://earthobservatory.nasa.gov/Features/WorldOfChange/hobet.php.

37. Emily Bernhardt and Margaret Palmer, "Environmental Costs of Mountaintop Mining Valley Fill Operations for Aquatic Ecosystems of the Central Appalachians," *Annals of the New York Academy of Sciences* 1223 (2011): 39–57, http://palmerlab.umd.edu/Bernhard_and_Palmer_2011.pdf.

38. John McQuaid, "Mining the Mountains," *Smithsonian*, January 2009, http://www.smithsonianmag.com/ecocenter-energy/mining-the-mountains-130454620/.

39. Ibid.

40. Ibid.

41. 4-methylcyclohexane methanol is an organic solvent used to wash the impurities from coal before it is burned to generate power. Little is known about its specific effects on human health; it is an industrial chemical, so its health effects have not been extensively studied. It is associated with eye and skin irritation, itching or rashes. High or prolonged exposure may cause nausea, vomiting, dizziness, headaches, diarrhea, and other symptoms. For specific information about this and other chemicals found in the Elk River spill, refer to C. Adams, A. Whelton, and J. Rosen, "Health Effects for Chemicals in 2014 West Virginia Chemical Release: Crude MCHM Compounds, PPH and DiPPH," *West Virginia Testing Assessment Project Literature Review*, March 17, 2014, http://www.dhsem.wv.gov/WVTAP/test-results/Documents/POSTED_WVTAP%20Health%20EFFECTs%20Lit%20Review%20v1.5%20031714.pdf.

42. Catie Talarski, "Update on West Virginia's Elk River Chemical Spill," NPR News, April 7, 2014, http://wnpr.org/post/update-west-virginias-elk-river-chemical-spill.

43. Alexandra Berzon and Kris Maher, "West Virginia Chemical-Spill Site Avoided Broad Regulatory Scrutiny," *Wall Street Journal*, January 13, 2014, http://online.wsj.com/news/articles/SB10001424052702303819704579317062273564766.

44. Claudia Copeland, "Mountaintop Mining: Background on Current Controversies," Congressional Research Service, December 2, 2013, http://www.fas.org/sgp/crs/misc/RS21421.pdf.

45. "How Coal Works," Union of Concerned Scientists, http://www.ucsusa.org/clean_energy/coalvswind/brief_coal.html.

46. Gregory Johnson and Scott Stephens, "Wind Power and Biofuels: A Green Dilemma for Wildlife Conservation," in *Energy Development and Wildlife Conservation in Western North America*, ed. David Naugle (Washington, DC: Island Press, 2011), 131–55.

47. "Elwha River Restoration," National Park Service, http://www.nps.gov/olym/naturescience/elwha-ecosystem-restoration.htm.

6. Weighing Significant Impacts, True Costs, and Mitigation

1. National Environmental Policy Act, 43 FR 56003, November 29, 1978; 44 FR 874, January 3, 1979, http://ceq.hss.doe.gov/nepa/regs/nepa/nepaeqia.htm.

2. Ibid.

3. "617: State Environmental Quality Review," 617.7 Determining Significance, New York State Department of Environmental Conservation, http://www.dec.ny.gov/regs/4490.html#18103.

4. California Environmental Quality Act, California Code of Regulations, chapter 3, title 14, appendix G, California Natural Resources Agency, http://resources.ca.gov/ceqa/docs/2014_CEQA_Statutes_and_Guidelines.pdf.

5. Rob Alkemade et al., "GLOBIO3: A Framework to Investigate Options for Reducing Global Terrestrial Biodiversity Loss," *Ecosystems* 12 (2009).

6. Christina Kennedy, Jessica Wilkinson, and Jennifer Balch, *Conservation Thresholds for Land Use Planners* (Washington, DC: Environmental Law Institute, 2003).

7. Michael T. Barbour et al., *Rapid Bioassessment Protocols for Use in Streams and Wadeable Rivers: Periphyton, Benthic Macroinvertebrates and Fish*, 2nd ed., EPA 841-B-99-002 (Washington, DC: US Environmental Protection Agency, 1999).

8. Thomas Schueler, Lisa Fraley-McNeal, and Karen Cappiella, "Is Impervious Cover Still Important? Review of Recent Research," *Journal of Hydrologic Engineering* 14 (2009): 309.

9. Derek Booth, "Forest Cover, Impervious-Surface Area, and the Mitigation of Urbanization Impacts in King County, Washington," University of Washington Department of Civil and Environmental Engineering, 2000, http://hdl.handle.net/1773/19552; Scott J. Goetz et al., "IKONOS Imagery for Resource Management: Tree Cover, Impervious Surfaces, and Riparian Buffer Analysis in the Mid-Atlantic Region," *Remote Sensing of Environment* 88 (2003): 195–208; C. Ernst, *Protecting the Source* (San Francisco: Trust for Public Land, 2004), 21.

10. "Watershed Forestry Resource Guide," Center for Watershed Protection and US Forest Service—Northeastern Area State & Private Forestry, 2008, http://www.forestsforwatersheds.org/urban-watershed-forestry/.

11. Kennedy, Wilkinson, and Balch, *Conservation Thresholds*; Environmental Law Institute, *Planner's Guide to Wetland Buffers for Local Governments* (Washington, DC: Environmental Law Institute, 2008), www.elistore.org/Data/products/d18. 01.pdf; Gary Bentrup, *Conservation Buffers: Design Guidelines for Buffers, Corridors, and Greenways*, Gen. Tech. Rep. SRS-109 (Asheville, NC: US Department of Agriculture, Forest Service, 2008).

12. Karen Cappiella, Anne Kitchell, and Tom Schueler, *Using Local Watershed Plans to Protect Wetlands* (Ellicott City, MD: Center for Watershed Protection, 2007); Karen Cappiella and Lisa Fraley-McNeal, *The Importance of Protecting Vulnerable Streams and Wetlands at the Local Level* (Ellicott City, MD: Center for Watershed Protection, 2007).

13. Map Service Center, Federal Emergency Management Agency, https://msc.fema.gov/webapp/wcs/stores/servlet/FemaWelcomeView?storeId=10001&catalogId=10001&langId=-1.

14. "Water Resources of the United States: USGS Flood Information," US Geological Survey, http://water.usgs.gov/floods/.

15. "Drinking Water Contaminants: List of Contaminants and Their MCLs," US Environmental Protection Agency, June 3, 2013, http://water.epa.gov/drink/contaminants/#List.

16. Renee Sharp and J. Paul Pestano, *Water Treatment Contaminants: Forgotten Toxics in American Water* (Washington, DC: Environmental Working Group, February 2013), http://static. ewg.org/reports/2013/water_filters/2013_tap_water_report_final.pdf; "Endocrine Disruption," Endocrine Disruption Exchange (TEDX), 2014, http://endocrinedisruption.org/endocrine-disruption/introduction/overview.

17. Massachusetts Office of Environmental Affairs, *Biomap: Guiding Land Conservation for Biodiversity in Massachusetts* (Boston: Massachusetts Division of Fisheries and Wildlife, 2001).

18. Chicago Region Biodiversity Council, *Biodiversity Recovery Plan* (Chicago: Chicago Region Biodiversity Council, 1999).

19. Danielle LaBruna and Michael W. Klemens, *Northern Wallkill Biodiversity Plan: Balancing Development and Environmental Stewardship in the Hudson River Estuary Watershed*, MCA Technical Paper No. 13 (New York: Wildlife Conservation Society, 2007).

20. Environment Canada, *How Much Habitat Is Enough? A Framework for Guiding Habitat Rehabilitation in Great Lakes Areas of Concern* (Downsview, ON: Canadian Wildlife Service, 2004), www.on.ec.gc.ca/wildlife/publications-e.html.

21. Thomas Winter et al., *Ground Water and Surface Water: A Single Resource*, US Geological Survey Circular 1139 (Denver: US Geological Survey, 1998); William Alley, Thomas Reilly, and O. Lehn Franke, *Sustainability of Groundwater Resources*, US Geological Survey Circular 1186 (Denver: US Geological Survey, 1999).

22. Russell Urban-Mead, *Dutchess County Aquifer Recharge Rates and Sustainable Septic System Density Recommendations* (Poughkeepsie, NY: Chazen Cos., 2006).

23. Chesapeake Bay Foundation, *State of the Bay Report 2012* (Annapolis, MD: Chesapeake Bay Foundation, 2012), 5, http://www.cbf.org/about-the-bay/state-of-the-bay/2012-report.

24. "40 CFR 1508.20—Mitigation," Cornell University Law School, Legal Information Institute, http://www.law.cornell.edu/cfr/text/40/1508.20.

25. US Fish and Wildlife Service, "Guidance Regarding Wetland Activities and Platte River Basin Depletions," US Fish and Wildlife Service Mountain–Prairie Region, September 4, 2008, http://www.fws.gov/platteriver/Documents/platte%20wetland%20guidance%204Sep2008.pdf.

26. Hye Kwon, Rebecca Winer, and Thomas Schueler, "Eight Tools of Watershed Protection in Developing Areas," Watershed Academy Web, US Environmental Protection Agency, http://cfpub.epa.gov/watertrain/pdf/modules/new_eighttools.pdf.

27. Gary Bentrup and J. Chris Hoag, *The Practical Streambank Bioengineering Guide* (Washington, DC: US Department of Agriculture, 1998).

28. Joel Salatin, "Small Steps Can Change Our World," *Mother Earth News*, June/July 2014, 73–75.

29. New York State Department of Health, *A Public Health Review of High Volume Hydraulic Hydrofracturing for Shale Gas Development*, December 2014, p. 12, http://www.health.ny.gov/press/reports/docs/high_volume_hydraulic_fracturing.pdf.

7. Overcoming Obstacles to Water Protection

1. Matthew McDonald, commencement address, State University of New York at Plattsburgh, May 17, 2014.

2. Hye Kwon, Rebecca Winer, and Thomas Schueler, "Eight Tools of Watershed Protection in Developing Areas," Watershed Academy Web, US Environmental Protection Agency, http://cfpub.epa.gov/watertrain/pdf/modules/new_eighttools.pdf.

3. Marla J. Stelk and Jeanne Christie, *Ecosystem Service Valuation for Wetland Restoration: What It Is, How to Do It, and Best Practice Recommendations* (Windham, ME: Association of State Wetland Managers, 2014).

4. Carolyn Thompson, "Lawsuits: Love Canal Still Oozes 35 Years Later," *USA Today*, November 2, 2013, http://www.usatoday.com/story/money/business/2013/11/02/suits-claim-love-canal-still-oozing-35-years-later/3384259/.

5. Edward J. Sullivan, "A Brief History of the Takings Clause," Washington University School of Law, St. Louis, http://landuselaw.wustl.edu/Articles/Brief_Hx_Taking.htm.

6. Florence Williams, "The Shovel Rebellion," *Mother Jones*, January/February 2001, http://www.motherjones.com/politics/2001/01/shovel-rebellion.

8. Strategies for Action

1. "The Story of Dryden: The Town That Fought Fracking (and Is Winning)," Earthjustice, http://earthjustice.org/features/the-story-of-dryden-the-town-that-fought-fracking-and-is-winning.

2. "Court Rules That New York Towns Can Ban Fracking," *EcoWatch*, June 30, 2014, http://ecowatch.com/2014/06/30/new-york-towns-ban-fracking/2/.

3. Susan Sharon, "Maine City Council Votes to Keep Tar Sands out of Its Port," NPR, July 22, 2014, http://www.npr.org/2014/07/22/334074055/maine-city-council-votes-to-keep-tar-sands-out-of-its-port.

4. "2010 National Land Trust Census," Land Trust Alliance, https://www.landtrustalliance.org/land-trusts/land-trust-census.

5. Clear Skies Over Orangeville, http://www.csoo.info/default.html.

6. ROSA-4-Rockland, http://rosa4rockland.org.

7. "We are ROSA: Community Support," http://rosa4rockland.org/rosa-101/we-are-rosa/.

8. Henry M. Robert, *Robert's Rules of Order* (New York: HarperCollins, 1990), 640.

9. Ibid., 540.

10. Bill Raney, president of the West Virginia Coal Association, quoted in "Mining the Mountains," by John McQuaid, *Smithsonian,* January 2009, http://www.smithsonianmag.com/ecocenter-energy/mining-the-mountains-130454620/?no-ist=&page=4.

11. "Laws and Executive Orders," US Environmental Protection Agency, http://www2.epa.gov/laws-regulations/laws-and-executive-orders.

12. "Clean Water Act," US Environmental Protection Agency, http://www.epa.gov/oecaagct/lcwa.html.

13. "Buffer Zone Restoration Guidelines," Wellesley, MA, Wetland Protection Committee, http://www.wellesleyma.gov/pages/wellesleyma_nrc/wetlands/bz%20guidelines, and http://www.sherbornma.org/Pages/SherbornMA_Conservation/Buffer%20Zone%20Restoration%20Guidelines.pdf.

14. Omega Institute, http://www.eomega.org/omega-in-action/key-initiatives/omega-center-for-sustainable-living.

15. "Rochesterians Concerned about Unsafe Shale-Gas Extraction," *No Frack Almanac,* http://www.r-cause.net/no-frack-almanac.html.

Conclusion

1. National Research Council, *Climate Stabilization Targets: Emissions, Concentrations, and Impacts over Decades to Millennia* (Washington, DC: National Academies Press, 2011).

2. World Wildlife Fund, *Zoological Society of London and Global Footprint Network, 2010 and Beyond: Rising to the Biodiversity Challenge,* 2004, http://www.wwf.org.uk/filelibrary/pdf/2010_and_beyond.pdf.

3. Reed Noss, Edward T. LaRoe III, and J. Michael Scott, *Endangered Ecosystems of the United States: A Preliminary Assessment of Loss and Degradation,* 1995, http://noss.cos.ucf.edu/papers/Noss%20et%20al%201995.pdf.

4. Carlos Corvalan et al., *Millennium Ecosystem Assessment: Ecosystems and Human Well-Being; Synthesis* (Geneva, Switzerland: World Health Organization, 2005).

5. "Marjory Stoneman Douglas: The Grande Dame of the Everglades," Women in Florida History, April 12, 2012, http://womeninfloridahistory.wordpress.com/2012/04/12/marjory-stoneman-douglas-the-grande-dame-of-the-everglades/.

6. Jon Pareles, "Pete Seeger, Champion of Folk Music and Social Change, Dies at 94," *New York Times,* January 28, 2014, http://www.nytimes.com/2014/01/29/arts/music/pete-seeger-songwriter-and-champion-of-folk-music-dies-at-94.html?_r=0.

7. National Conservation Easement Data Base, updated September 2013, http://conservationeasement.us.

8. "Quotes from Our Native Past," ancient Indian proverbs, http://www.ilhawaii.net/~stony/quotes.html.

9. Judith Burns, "Children Urged to Put Away Screens and Play Outside," BBC News, October 25, 2013, BBC%20News%20%20Children%20urged%20to%20put%20away%20screens%20and%20play%20outside.webarchive.

10. Richard Louv, *Last Child in the Woods* (Chapel Hill, NC: Algonquin Books, 2006), 2.

11. Ibid., 158.

12. "Today's Worst Watershed Stresses May Become the New Normal," University of Colorado at Boulder, September 18, 2013, http://www.colorado.edu/news/releases/2013/09/18/today%E2%80%99s-worst-watershed-stresses-may-become-new-normal-study-finds#sthash.7TwVKOVg.dpuf.

13. "Study Reveals Parched U.S. West Using Up Underground Water," *NASA News,* July 24, 2014, http://www.nasa.gov/press/2014/july/satellite-study-reveals-parched-us-west-using-up-underground-water/#.U9aN3FaBX1p.

14. Caryn Ernst, *Protecting the Source: Land Conservation and the Future of America's Drinking Water* (San Francisco: Trust for Public Land, 2004).

Appendix: Stream and Wetland Checklists

1. A. J. K. Calhoun and M. W. Klemens, *Best Development Practices: Conserving Pool-Breeding Amphibians in Residential and Commercial Developments in the Northeastern United States*, MCA Technical Paper No. 5 (New York: Metropolitan Conservation Alliance, Wildlife Conservation Society, 2002).

selected bibliography

Alley, William, T. Reilly, and O. Franke. *Sustainability of Ground-Water Resources*. US Geological Survey Circular 1186. Reston, VA: Government Printing Office, 1999. http://pubs.usgs.gov/circ/circ1186/html/gen_facts.html.

Barbour, Michael T., Jereon Gerritsen, Blaine D. Snyder, and James B. Stribling. *Rapid Bioassessment Protocols for Use in Streams and Wadeable Rivers: Periphyton, Benthic Macroinvertebrates and Fish*. 2nd edition. EPA 841-B-99–002. Washington, DC: US Environmental Protection Agency, 1999.

Barlow, Paul, and Stanley Leake. *Streamflow Depletion by Wells—Understanding and Managing the Effects of Groundwater Pumping on Streamflow*. US Geological Survey Circular 1376. Reston, VA: Government Printing Office, 2012. http://pubs.usgs.gov/circ/1376/.

Bentrup, Gary. *Conservation Buffers: Design Guidelines for Buffers, Corridors, and Greenways*. Gen. Tech. Rep. SRS-109. Asheville, NC: Department of Agriculture, Forest Service, Southern Research Station, 2008.

Calhoun, Aram, and Michael W. Klemens. *Best Development Practices: Conserving Pool-Breeding Amphibians in Residential and Commercial Developments in the Northeastern United States*. MCA Technical Paper No. 5. New York: Wildlife Conservation Society, 2002.

Cappiella, Karen, and L. Fraley-McNeal. *The Importance of Protecting Vulnerable Streams and Wetlands at the Local Level*. Ellicott City, MD: Center for Watershed Protection, 2007.

Cappiella, Karen, Anne Kitchell, and Tom Schueler. *Using Local Watershed Plans to Protect Wetlands*. Ellicott City, MD: Center for Watershed Protection, 2006.

Cappiella, Karen, Tom Schueler, Julie Tasillo, and Tiffany Wright. *Adapting Watershed Tools to Protect Wetlands*. Ellicott City, MD: Center for Watershed Protection, 2005.

Chesapeake Bay Foundation. *State of the Bay Report 2012*. Annapolis, MD: Chesapeake Bay Foundation, 2012. http://www.cbf.org/about-the-bay/state-of-the-bay/2012-reporthttp://www.cbf.org/about-the-bay/state-of-the-bay/2012-report.

Corvalan, Carlos, et al. *Millennium Ecosystem Assessment: Ecosystems and Human Well-Being; Synthesis*. Geneva, Switzerland: World Health Organization, 2005.

Environment Canada. *How Much Habitat Is Enough? A Framework for Guiding Habitat Rehabilitation in Great Lakes Areas of Concern*. Downsview, ON: Canadian Wildlife Service, 2004. www.on.ec.gc.ca/wildlife/publications-e.html.

Ernst, C., R. Gullick, and K. Nixon. "Protecting the Source: Conserving Forests to Protect Water." *Opflow* 30 (2004).

Guth, Joseph. "Cumulative Impacts: Death-Knell for Cost-Benefit Analysis in Environmental Decisions. *Barry Law Review* 11 (Fall 2008).

Harker, Donald, and Elizabeth Natter. *Where We Live: A Citizen's Guide to Conducting a Community Environmental Inventory.* Washington, DC: Island Press, 1995.

Johnson, Elizabeth, and Michael Klemens, eds. *Nature in Fragments: The Legacy of Sprawl.* New York: Columbia University Press, 2005.

Kennedy, Christina, Jessica Wilkinson, and Jennifer Balch. *Conservation Thresholds for Land Use Planners.* Washington, DC: Environmental Law Institute, 2003.

Kiviat, Erik, and Gretchen Stevens. *Biodiversity Assessment Manual for the Hudson River Estuary Corridor.* Annandale, NY: Hudsonia Ltd., 2001.

Klein, Naomi. *This Changes Everything.* New York: Simon & Schuster, 2014.

Klemens, Michael W., Marjorie Shansky, and Henry Gruner. *From Planning to Action: Biodiversity Conservation in Connecticut Towns.* MCA Technical Paper No. 10. New York: Wildlife Conservation Society, 2006.

Kwon, Hye, Rebecca Winer, and Thomas Schueler. "Eight Tools of Watershed Protection in Developing Areas." Watershed Academy Web, US Environmental Protection Agency. http://cfpub.epa.gov/watertrain/pdf/modules/new_eighttools.pdf.

LaBruna, Danielle, and Michael W. Klemens. *Northern Wallkill Biodiversity Plan: Balancing Development and Environmental Stewardship in the Hudson River Valley Watershed.* MCA Technical Paper 13. New York: Wildlife Conservation Society, 2007.

Leopold, Aldo. *A Sand County Almanac.* New York: Oxford University Press, 1966.

Louv, Richard. *Last Child in the Woods.* Chapel Hill, NC: Algonquin Books, 2006.

McElfish, James, Rebecca Kihslinger, and Sandra Nichols. *Planner's Guide to Wetland Buffers for Local Governments.* Washington, DC: Environmental Law Institute, 2008.

McQuaid, John. "Mining the Mountain." *Smithsonian Magazine,* January 2009.

Millennium Ecosystem Assessment Board. *Living beyond Our Means: Natural Assets and Human Well-Being.* United Nations Millennium Ecosystem Assessment, 2005. http://www.maweb.org/en/BoardStatement.aspx.

Robert, Henry M. *Robert's Rules of Order.* 1990 edition. New York: HarperCollins.

Schueler, Thomas. "Urban Pesticides: From the Lawn to the Stream." *Watershed Protection Techniques* 2 (1995): 247–53.

Stelk, Marla, and Jeanne Christie. *Ecosystem Service Valuation for Wetland Restoration: What It Is, How to Do It, and Best Practice Recommendations.* Windham, ME: Association of State Wetland Managers, 2014.

Tallamy, Douglas. *Bringing Nature Home.* Portland, OR: Timber Press, 2007.

Tiner, Ralph. *Wetland Indicators: A Guide to Wetland Identification, Delineation, Classification and Mapping.* Boca Raton, FL: CRC Press, 1999.

Urban-Mead, Russell. *Dutchess County Aquifer Recharge Rates and Sustainable Septic System Density Recommendations.* Poughkeepsie, NY: Chazen Cos., 2006.

"Watershed Forestry Resource Guide." Center for Watershed Protection and US Forest Service—Northeastern Area State & Private Forestry, 2008. http://www.forestsforwatersheds.org/urban-watershed-forestry/.

Weidenhof, Emily, Vivian Ngo, and Richard Gonzalez. *Hancock and the Marcellus Shale: Visioning the Impacts of Natural Gas Extraction along the Upper Delaware.* New York: Open Space Institute, 2009.

Winter, Thomas, Judson W. Harvey, O. Lehn Franke, and William M. Alley. *Ground Water and Surface Water: A Single Resource.* US Geological Survey Circular 1139. Reston, VA: Government Printing Office, 1999.

Wright, Tiffany, Jennifer Tomlinson, Tom Shueler, Karen Cappiella, Anne Kitchell, and Dave Hirshman. *Direct and Indirect Effects of Urbanization on Wetland Quality.* Ellicott City, MD: Center for Watershed Protection, 2006.

index